T0135759

INAUGURAL - DISSERTATION

zur

Erlangung der Doktorwürde

der

Naturwissenschaftlich-Mathematischen Gesamtfakultät

der

Ruprecht - Karls - Universität

Heidelberg

vorgelegt von

Diplom-Informatikerin der Medizin Nathalie Harder

aus Forchheim

Tag der mündlichen Prüfung: 20.04.2010

Bibliografische Information der Deutschen Nationalbibliothek

Die Deutsche Nationalbibliothek verzeichnet diese Publikation in der
Deutschen Nationalbibliografie; detaillierte bibliografische Daten sind
im Internet über http://dnb.d-nb.de abrufbar.

ISBN 978-3-8325-2530-9

Logos Verlag Berlin GmbH
Comeniushof, Gubener Str. 47,
10243 Berlin
Tel.: +49 (0)30 42 85 10 90
Fax: +49 (0)30 42 85 10 92
INTERNET: http://www.logos-verlag.de

Automatic Cell Cycle Analysis Based on Live Cell Fluorescence Microscopy Image Sequences

1. Gutachter: Prof. Dr. Roland Eils
2. Gutachter: Prof. Dr.-Ing. Hartmut Dickhaus

Zusammenfassung

Diese Arbeit befasst sich mit der automatischen Auswertung großer Mengen von 2D und 3D fluoreszenzmikroskopischer Bildsequenzen. In biologischen Hochdurchsatzexperimenten können mittels automatisierter Laborsysteme große Bilddatenmengen mit hohem Informationsgehalt aufgenommen werden. Die Auswertung dieser großen Datenmengen stellt jedoch oft ein Problem dar, da manuelle Auswertung kaum zu bewältigen ist und existierende automatische Bildanalysesysteme für die oft sehr spezifischen Anwendungen meist nicht direkt anwendbar sind. In dieser Arbeit stellen wir ein automatisches Bildanalysesystem zur detaillierten Zellzyklusanalyse in Fluoreszenzmikroskopiebildern lebender Zellen vor. Unser Bildanalysesystem bestimmt robust und automatisch die Dauer der einzelnen Zellzyklusphasen und ermöglicht somit die Auswertung großangelegter Screening Experimente zur systematischen Erforschung der Genregulation der Zellteilung.

Das hier entwickelte Verfahren kombiniert verschiedene Bildverarbeitungsmethoden in einem komplexen Ablauf, der eine automatische Auswertung von 3D Bildsequenzen von konfokalen Fluoreszenzmikroskopen ermöglicht. Unser Bildanalysesystem beinhaltet Methoden für Segmentierung, Tracking, Merkmalsextraktion, Klassifikation in Zellzyklusphasen, und Phasensequenzprüfung. Für die schnelle und robuste Segmentierung von Zellkernen wurde ein regionenbasiertes Schwellenwertverfahren entwickelt. Das Segmentierungsverfahren ermöglicht insbesondere die korrekte Segmentierung morphologisch ungewöhnlicher Zellkerne, z.B. mit anhängenden, kontrastarmen Mikrozellkernen, oder abseitsliegenden Kernbruchstücken. Die Segmentierungsgenauigkeit wurde anhand von realen Bilddaten umfassend validiert. Zur Verfolgung lebender Zellen über die Zeit haben wir ein Merkmalspunkt-basiertes Trackingverfahren mit einer neuen Methode zur Mitosedetektion erweitert, die eine korrekte Verfolgung von Zellteilungen erlaubt. Diese neue Methode zur Mitosedetektion basiert auf einer Likelihood Funktion, in die verschiedene morphologische und topologische Eigenschaften von Mutter- und Tochterzellkernen einfließen. Die Genauigkeit unseres Trackingverfahrens wurde basierend auf realen Bilddaten evaluiert. Um eine automatische Klassifikation der Zellkerne in Zellzyklusphasen zu ermöglichen, wurden die Zellkerne zunächst durch eine Reihe geeigneter Bildmerkmale charakterisiert. Diese Zusammenstellung von Merkmalen wurde durch neue, dynamische Bildmerkmalen erweitert, was zu einer deutlichen Verbesserung der Klassifikationsgenauigkeit führte. Des Weiteren wurden verschiedene Merkmalskombinationen und Merkmalsreduktionsverfahren hinsichtlich der resultierenden Klassifikationsgenauigkeiten untersucht. Für die automatische Klassifikation verwenden wir Support-Vector-Maschinen, doch auch die Eignung anderer überwachter und nicht-überwachter Lernverfahren wurde systematisch untersucht. Bei der Klassifikation der Zellkerne unterscheiden wir bis zu 13 Klassen, darunter sieben reguläre Zellzyklusphasen und sechs morphologische Phänotypklassen. Eine Vielzahl unterschiedlicher Experimente wurde durchgeführt, die eine durchgängig hohe Klassifikationsgenauigkeit bestätigten. Zur Bestimmung der Zellzyklusphasendauern basierend auf den Tracking- und Klassifikationsergebnissen haben wir eine neue Methode zur Phasensequenzanalyse entwickelt. Diese Methode basiert auf einem endlichen Automaten, welcher ausschließlich biologisch sinnvolle Phasensequenzen akzeptiert. Inkonsistenzen der Phasensequenz werden detektiert und korrigiert, und Phasendauern werden gemessen.

Im Rahmen dieser Arbeit haben wir unser System auf eine sehr große Anzahl realer Bilder von vier verschiedenen Screening Experimenten angewendet. Dabei wurden insgesamt mehr als 1000 Bildsequenzen ausgewertet, von denen jede ca. 100 bis 200 Zeitschritte umfasst. Die experimentelle Auswertung zeigte, dass unser System verschiedene Problemstellungen robust und mit hoher Genauigkeit löst: (1) Klassifikation und Quantifizierung von Zellkernen in verschiedenen Zellzyklusphasen, (2) Bestimmung der Zellzyklusphasenlängen, und (3) Klassifikation und Quantifizierung morphologischer Phänotypklassen. Zudem enthalten die berechneten Trackingergebnisse weitere wertvolle Informationen die bisher noch nicht vollständig ausgewertet wurden und in zukünftigen Arbeiten zur detaillierten Bewegungsanalyse verwendet werden können. Um die breite Anwendbarkeit unseres Systems unter verschiedenen biologischen Rahmenbedingungen zu demonstrieren, wurden bei der experimentellen Evaluierung Bilder verschiedener Zelllinien und Bilder von verschiedenen Screening-Mikroskopen analysiert.

Abstract

In this thesis, we address the problem of automatic analysis of large sets of 2D and 3D fluorescence microscopy image sequences. Automated screening platforms allow biologists to acquire large amounts of image data with high information content. However, analyzing this data is often problematic since manual analysis is not feasible, and suitable automated image analysis approaches are generally not available off-the-shelf for specific analysis tasks. Here, we present an automatic approach for detailed cell cycle analysis based on live cell fluorescence microscopy image sequences. In particular, our approach enables robust automatic determination of cell cycle phase durations which allows automatically analyzing large-scale screening experiments to study mitotic gene regulation.

In this work, we developed a complex image analysis workflow which evaluates 3D image sequences from confocal fluorescence microscopy. Our workflow comprises methods for segmentation, tracking, feature extraction, classification into cell cycle phases, and phase sequence parsing. For fast and robust segmentation of cell nuclei we developed an approach based on region-adaptive thresholding. In particular, this approach copes with morphologically highly abnormal cell nuclei, e.g., nuclei with attached dim micronuclei or detached nucleus parts. We validated the accuracy of our segmentation approach based on real image data. For tracking of live cell nuclei we combined a feature-point tracking algorithm with a newly developed approach for robust mitosis detection in order to track cell divisions properly. The mitosis detection approach is based on a likelihood function taking into account different morphological and topological properties of mother and daughter cell nuclei. The performance of our tracking approach was evaluated based on real data. To enable classification of cell nuclei into cell cycle phases we characterize the nuclei with a set of descriptive image features. We enhanced this set of static features by dynamic image features which significantly improved the classification accuracy. Furthermore, we studied the performance for differently composed feature sets and feature reduction strategies. For cell classification we use support vector machines, however, also other supervised and unsupervised learning methods were studied.

We classify cell nuclei into up to 13 classes, comprising seven cell cycle phases and six morphological phenotype classes. A variety of different experiments was performed, resulting in overall high classification accuracies. To determine cell cycle phase durations based on the tracking and classification results we developed an error-correcting phase sequence parser. This approach is based on a finite state machine which accepts only biologically plausible sequences of cell cycle phases. Inconsistencies within phase sequences are detected, corrected, and phase durations are measured.

Our approach was applied to a large number of real images from four different screens. In total, we analyzed more than 1000 image sequences with around 100 to 200 time steps each. The experimental evaluation showed that our approach robustly and accurately solves different tasks: (1) classification and quantification of cell nuclei into different cell cycle phases, (2) determination of cell cycle phase durations, and (3) classification and quantification of morphological phenotype classes. Furthermore, the tracking results provide additional information for detailed motion analysis of cells which has not been fully exploited here and could serve as a basis in future studies. In our experiments we also demonstrated the applicability of our approach for different biological settings by analyzing images of different cell lines and images of different screening microscopes.

Acknowledgement

I would like to sincerely thank all people who supported me during my thesis work!

First, I would like to thank Roland Eils for giving me the opportunity to work on this interesting project, for his confidence, and for always supporting my work.

Moreover, many thanks to Karl Rohr for his excellent advice, guidance, and support, and for all the time he invested to increase the quality of our papers and of this thesis. *Thanks a lot!*

Thanks also to Prof. Dickhaus for agreeing to be the second referee of this thesis.

Next, I would like to thank our cooperation partners at EMBL, Jan Ellenberg, Felipe, and Annelie for providing me with lots of interesting biological background knowledge, for the good cooperation, and for all the fascinating images.

Many thanks to all colleagues involved in collaborative work for the fruitful cooperations, i.e. our guest scientist Vassili Kovalev, the group of Rainer König, the group of Guillermo Marcus and Prof. Männer in Mannheim, Il-Han and Petr. I also would like to thank all my internship students for their various types of input as well as William for contributing with his bachelor thesis. In addition, I am grateful for the previous work I could build on, provided by former group members such as Christian Conrad, Patrick Warnat, and Christian Bacher.

For providing the computational environment to accomplish this work I would like to thank Karlheinz, Rolf, and Peter as well as the IT support team at Bioquant.

My colleagues Stefan and William I would like to thank for reading parts of my thesis and for being enjoyable office mates. Furthermore, I thank all my colleagues at the iBioS division and, in particular, all former and current members of the Biomedical Computer Vision group for the pleasant working atmosphere.

Finally, special thanks go to my parents who supported me in all possible ways, and to Markus for always being there for me and enduring all the overtime I spent at work over the past years. *Thanks so much!*

List of publications

Conferences

- **N. Harder**, B. Neumann, M. Held, U. Liebel, H. Erfle, J. Ellenberg, R. Eils, and K. Rohr. Automated recognition of mitotic phenotypes in fluorescence microscopy images of human cells. In H. Handels, J. Ehrhardt, A. Horsch, H.-P. Meinzer, and T. Tolxdorff, editors, *Proc. Workshop Bildverarbeitung für die Medizin 2006 (BVM'06)*, Informatik aktuell, pages 206–210, Hamburg, Germany, Mar 19–21 2006. Springer-Verlag.

- **N. Harder**, B. Neumann, M. Held, U. Liebel, H. Erfle, J. Ellenberg, R. Eils, and K. Rohr. Automated recognition of mitotic patterns in fluorescence microscopy images of human cells. In J. Kovačević and E. Meijering, editors, *Proc. IEEE Internat. Symposium on Biomedical Imaging: From Nano to Macro (ISBI'2006)*, pages 1016–1019, Arlington, VA, USA, Apr 6–9 2006.

- V. Kovalev, **N. Harder**, B. Neumann, M. Held, U. Liebel, H. Erfle, J. Ellenberg, R. Eils, and K. Rohr. Feature selection for evaluating fluorescence microscopy images in genome-wide cell screens. In C. Schmid, S. Soatto, and C. Tomasi, editors, *Proc. IEEE Computer Society Conf. on Computer Vision and Pattern Recognition (CVPR'06)*, pages 276–283, New York, NY, USA, Jun 17–22 2006.

- **N. Harder**, F. Mora-Bermúdez, W.J. Godinez, J. Ellenberg, R. Eils, and K. Rohr. Automated analysis of the mitotic phases of human cells in 3D fluorescence microscopy image sequences. In R. Larsen, M. Nielsen, and J. Sporring, editors, *Proc. 9th Internat. Conf. on Medical Image Computing and Computer-Assisted Intervention (MICCAI'2006)*, volume 4190 of *LNCS*, pages 840–848, Copenhagen, DK, Oct 1–6 2006. Springer-Verlag.

- **N. Harder**, F. Mora-Bermúdez, W.J. Godinez, J. Ellenberg, R. Eils, and K. Rohr. Automated analysis of the mitotic phases of human cells in 3D fluorescence microscopy image sequences. In D.N. Metaxas, R.T. Whitaker, J. Rittscher, and

T.B. Sebastian, editors, *Proc. MICCAI'06 Workshop Microscopic Image Analysis with Applications in Biology (MIAAB'2006)*, Copenhagen, DK, Oct 5 2006.

- I.-H. Kim, W.J. Godinez, **N. Harder**, F. Mora-Bermúdez, J. Ellenberg, R. Eils, and K. Rohr. Compensation of global movement for improved tracking of cells in time-lapse confocal microscopy image sequences. In J.P.W. Pluim and J.M. Reinhardt, editors, *Medical Imaging 2007: Image Processing (MI'07), Proc. SPIE*, volume 6512, San Diego, CA, USA, Feb 17–22 2007.

- **N. Harder**, F. Mora-Bermúdez, W.J. Godinez, J. Ellenberg, R. Eils, and K. Rohr. Determination of mitotic delays in 3D fluorescence microscopy images of human cells using an error-correcting finite state machine. In A. Horsch, T.M. Deserno, H. Handels, H.-P. Meinzer, and T.Tolxdorff, editors, *Proc. Workshop Bildverarbeitung für die Medizin 2007 (BVM'07)*, Informatik aktuell, pages 242–246, Munich, Germany, Mar 25–27 2007. Springer-Verlag.

- **N. Harder**, F. Mora-Bermúdez, W.J. Godinez, J. Ellenberg, R. Eils, and K. Rohr. Determination of mitotic delays in 3D fluorescence microscopy images of human cells using an error-correcting finite state machine. In J. Fessler and T. Denney, editors, *Proc. IEEE Internat. Symposium on Biomedical Imaging: From Nano to Macro (ISBI'2007)*, pages 1044–1047, Arlington, VA, USA, Apr 12–15 2007.

- **N. Harder**, F. Mora-Bermúdez, W.J. Godinez, J. Ellenberg, R. Eils, and K. Rohr. Automated analysis of mitotic cell nuclei in 3D fluorescence microscopy image sequences. In B.S. Manjunath, G. Danuser, and A. Carpenter, editors, *Workshop on Bio-Image Informatics: Biological Imaging, Computer Vision and Data Mining*, Center for Bio-Image Informatics, UCSB, Santa Barbara, CA, USA, Jan 17–18 2008.

- P. Matula, A. Kumar, I. Wörz, **N. Harder**, H. Erfle, R. Bartenschlager, R. Eils, and K. Rohr. Automated analysis of siRNA screens of cells infected by Hepatitis C and Dengue viruses based on immunofluorescence microscopy images. In X.P. Hu and A.V. Clough, editors, *Medical Imaging 2008: Physiology, Function, and Structure from Medical Images (MI'08), Proc. SPIE*, San Diego, CA, USA, Feb 16–21 2008.

- P. Matula, A. Kumar, I. Wörz, **N. Harder**, H. Erfle, R. Bartenschlager, R. Eils, and K. Rohr. Automated analysis of siRNA screens of virus infected cells based on immunofluorescence microscopy images. In T. Tolxdorff, J. Braun, T.M. Deserno, H. Handels, A. Horsch, and H.-P. Meinzer, editors, *Proc. Workshop*

Bildverarbeitung für die Medizin 2008 (BVM'08), Informatik aktuell, pages 453–457, Berlin, Germany, Apr 6–8 2008. Springer-Verlag.

- M. Gipp, G. Marcus, **N. Harder**, A. Suratanee, K. Rohr, R. König, and R. Männer. Accelerating the computation of Haralick's texture features using graphics processing units (GPUs). In *Proc. World Congress on Engineering 2008 (WCE'08), The 2008 Internat. Conf. of Parallel and Distributed Computing (ICPDC'08)*, pages 587–593, London, UK, Jul 2–4 2008. Newswood Limited, International Association of Engineers.

Book Chapters

- **N. Harder**, R. Eils, and K. Rohr. Automated classification of mitotic phenotypes of human cells using fluorescent proteins. In K.F. Sullivan, editor, *Fluorescent Proteins*, volume 85 of *Methods in Cell Biology*, pages 539–554. Academic Press, 2008.

- **N. Harder**, F. Mora-Bermúdez, W.J. Godinez, J. Ellenberg, R. Eils, and K. Rohr. Automated analysis of the mitotic phases of human cells in 3D fluorescence microscopy image sequences. In J. Rittscher, R. Machiraju, and S.T.C. Wong, editors, *Microscopic Image Analysis for Life Science Applications*, pages 283–293. Artech House, 2008.

Journals

- **N. Harder**, F. Mora-Bermúdez, W.J. Godinez, A. Wünsche, R. Eils, J. Ellenberg, and K. Rohr. Automatic analysis of dividing cells in live cell movies to detect mitotic delays and correlate phenotypes in time. *Genome Research*, 19(11):2113–2124, Nov 2009.

- M. Gipp, G. Marcus, **N. Harder**, A. Suratanee, K. Rohr, R. König, and R. Männer. Haralick's texture features computed by GPUs for biological applications. *IAENG International Journal of Computer Science*, 36(1), 2009.

Contents

Chapter 1

Introduction

1.1 Motivation

The need for image analysis methods in biological research has increased significantly in recent years. This is because large amounts of image data can be acquired in very short time frames using highly automated laboratories. At the same time, the costs for data storage have become low so that many laboratories have the possibility to perform large-scale experiments producing terabytes of data per day. However, to derive meaningful results, the acquired data has to be evaluated and quantified. Generally, this can be done either manually by human experts or automatically using computational methods. Manual data analysis is very time consuming when dealing with large data sets, and in addition, requires experienced experts. Also, the acquired image data is often high-dimensional, e.g., 3D image sequences with multiple color channels, which makes manual analysis even more difficult. To overcome this limitation, automatic or semi-automatic image analysis methods have to be developed that reach comparable accuracies as human experts.

This thesis is concerned with the development of approaches for the automatic evaluation of cell microscopy images from such large-scale experiments. In particular, we consider 2D and 3D image sequences which have been acquired in the framework of the EU project MitoCheck [5]. The goal of this project is to study the process of cell division (mitosis) in human cells at a molecular level. In this context, high-throughput screening experiments are performed to identify and characterize the genes that control cell division. The method of choice to study gene function is the technology of RNA interference (RNAi). With RNAi genes are systematically knocked down, i.e. they are no longer expressed, and thus the processes they normally regulate are disturbed. By monitoring the cellular processes that are affected by gene knockdown, one can deduce knowledge on the function of the respective

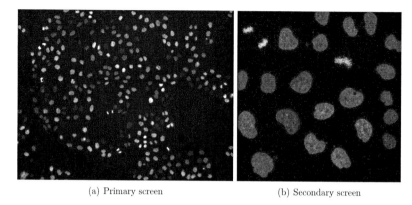

(a) Primary screen   (b) Secondary screen

Figure 1.1: Example images. (a) Typical image frame of a 2D image sequence from the primary screen, (b) maximum intensity projection of a typical image stack of a 3D image sequence from the secondary screen.

gene. To study the genes required for cell division, RNAi-treated cell cultures are observed over several hours and their mitotic behavior is analyzed. To this end, 2D and 3D image sequences are acquired using automated fluorescence microscopy at the European Molecular Biology Laboratory (EMBL) Heidelberg. Two typical images from the MitoCheck project are shown in Fig. 1.1. In this thesis, we developed an image analysis approach to automatically analyze the mitotic behavior of cell nuclei. The analysis tasks that have been addressed here are first, to quantify cell nuclei in different stages of the cell cycle, and second, to analyze the duration of the cell cycle phases automatically. To this end, we developed an image analysis workflow comprising methods for cell sementation, tracking, image feature extraction, classification, and mitotic phase sequence processing.

1.2   Biological and Biotechnical Background

In this chapter, we provide a brief summary of the biological and biotechnical background to illustrate the application context of our work, particularly for the biologically less-trained reader. First, we shortly describe the process of mitosis which is the biological target of all studies considered in this thesis. Next, we provide an overview of the applied imaging modalities: widefield and confocal fluorescence microscopy. Thereafter, we briefly explain the main idea of RNA interference for gene knockdown, followed by a short description of a typical workflow for automated high-throughput high-content experiments. Finally, we give an outline of the

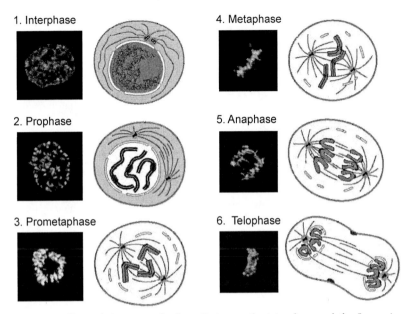

Figure 1.2: Example images and schematic images for interphase and the five main phases of mitosis: prophase, prometaphase, metaphase, anaphase, and telophase. For each phase a real image of the cell nucleus from a confocal fluorescence microscope is given on the left (HeLa (Kyoto) labeled with histone EGFP H2B), and a schematic sketch of the whole cell on the right (adapted from [9]).

most common tasks for automatic image analysis in this context. For more detailed information on these topics we refer to the standard cell biology literature (e.g., [9]).

1.2.1 Mitosis: the process of cell division

The cell cycle of a eukaryotic cell usually is subdivided into interphase and M-phase, where M-phase denotes the process of cell division. Interphase can be further subdivided into G1-, S-, and G2-phase. In S-phase the DNA is replicated before the next cell division can take place, while the G-phases denote so called gap phases in which other cellular growth processes take place. M-phase involves the division of the cell nucleus (mitosis) followed by the division of the cytoplasm (cytokinesis). Since in the following only nuclear division is considered, we focus on the process of mitosis.

Mitosis is typically defined as a strictly ordered sequence of five phases: prophase, prometaphase, metaphase, anaphase, and telophase. Figure 1.2 shows for each phase, on the left, a real image of the cell nucleus from a confocal fluorescence microscope, and on the right, a schematic sketch of the whole cell. The cycle starts with interphase, the phase before the actual start of mitosis. In interphase the chromatin is diffuse and relatively homogeneously distributed in the cell nucleus. Most of its life time a cell is in interphase, to grow and perform all normal cell functions, and to prepare for mitosis. The cell enters mitosis with prophase as the chromatin starts to condense into chromosomes. Each chromosome consists of two identical sister chromatides which are connected via a kinetochore. During prophase also the mitotic spindle begins to form, as the two spindle poles move to the cell poles. In prometaphase, the nuclear envelope breaks down abruptly which allows the dynamic fibers forming the mitotic spindle to enter the nuclear region and attach to the kinetochores. The spindle fibers then start to arrange the chromosomes into one plane between the spindle poles: the metaphase plate. As soon as all chromosomes are located in the metaphase plate, metaphase is completed and anaphase is initiated. In early anaphase (or anaphase A) the kinetochores separate, which enables the spindle fibers to pull the chromosomes' sister chromatides to the opposite cell poles, and thus perform the actual chromosome segregation. In late anaphase (or anaphase B) the cell poles additionally move further apart from each other to increase the space between the forming daughter cell nuclei. In telophase the chromosome sets have arrived at the cell poles, new nuclear envelopes re-assemble around the daughter chromosome sets, and finally, the chromosomes decondense. Mitosis ends with the formation of two new daughter cell nuclei, and after cytokinesis the cell goes over to interphase again.

In conclusion, mitosis is an essential process for growth and regeneration in higher multicellular organisms. It is a very complex process which is mediated by a large variety of proteins, and consequently, is regulated by a high number of genes. Errors during mitosis mostly have severe consequences for the organism, as for example cancer manifests as a disturbed regulation of mitosis. However, up to now the details of mitosis regulation are not yet fully understood and systematic screening of mitosis-related genes helps to elucidate these processes.

1.2.2   Fluorescence microscopy

Fluorescence microscopy is an important imaging method which is often applied to observe dynamic processes in living cells as, for example, mitosis. This imaging technology allows visualizing specifically labelled cellular structures at high resolution. Fluorescence microscopy is based on the physical effect of fluorescence, i.e. fluorescent molecules (fluorophores) are excited with laser light of a certain wavelength, and consequently, emit light of an increased wavelength (i.e. decreased energy). Thus, a fluorescence microscope requires an excitation laser, followed by a set of filters which allow selecting the fluorophore-specific wavelength by blocking all other wavelengths. The transmitted light is directed to the fluorescently labeled sample where it excites the fluorophore. To detect the light emitted by the fluorophore, a second set of filters is required, which pass only the emitted wavelength and filter out the excitation light. A digital image can then be acquired based on the emitted light with a CCD (charge-coupled device) camera. Consequently, the fluorophore-specific excitation wavelength, as well as the emission wavelength, have to be known. To visualize cellular structures based on this effect, the fluorophores are attached to the structures of interest within the cell. A common labeling strategy is to couple the fluorescent molecules to antibody molecules which bind specifically to the structures of interest in the cell (e.g., DAPI). However, such fluorescent dyes have to be externally created and then inserted into the cell. A more elegant strategy is to use natural fluorescent proteins such as, e.g., GFP (green fluorescent protein, originating from jellyfish *A. victoria*), which can be produced by the cell itself. To this end, a specific gene sequence encoding the blueprint of the fluorescent protein has to be transfected into the cellular genome. The advantage of such fluorescent proteins is that living cells (stably expressing the fluorescent protein) can be visualized over a long time without disturbing them by staining. By inserting the fluorescent protein-encoding DNA sequence at the respective position of the genome, also fluorescent protein-fusion proteins can be created which behave like the original protein but are fluorescently labeled. This enables a wide range of applications in live cell imaging. Moreover, multiple different structures can be visualized at once using different fluorescent markers. The light emitted by the different fluorophores is acquired through the respective filters in distinct imaging channels.

An enhancement of the above described conventional (widefield) fluorescence microscope is the confocal laser scanning microscope, which provides an increased spatial resolution and enables 3D imaging. In comparison to a conventional fluorescence microscope, the confocal system does not illuminate the whole sample at once, but focuses on only one point. To this end, a pair of confocal pinholes are used, which on the illumination side, focus the laser on one point in the sample, and on the detection side, exclude the emitted light that does not come from the point of focus (see Fig. 1.3). A complete image section is collected by scanning the whole focal plane point-wise in a raster pattern. For this purpose, an oscillating mirror is applied to deflect the light beam between the objective and the dichroic mirror. A 3D image of an object can be created by scanning multiple focal planes in different

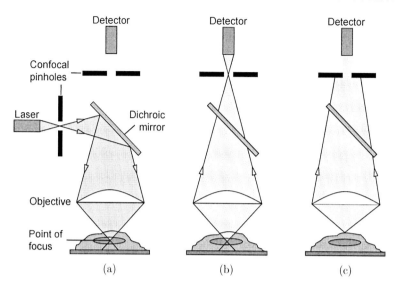

Figure 1.3: Basic concept of a confocal fluorescence microscope. The microscope can focus on one point in the 3D specimen using two confocal pinholes. The dichroic mirror reflects the excitation light and passes the emitted light. (a) Excitation light beam focused by the illumination pinhole. (b) Light emitted from the point of focus which passes the detection pinhole and reaches the detector. (c) Emitted out-of-focus light which is largely excluded from detection by the detection pinhole. (Adapted from [9].)

depths of the sample. Thus, confocal fluorescence microscopy allows acquiring 5D image data (i.e. space in x-, y-, and z-direction, time, and color channels).

1.2.3   Gene knockdown based on RNA interference

A common technique to study the function of a certain gene is to knock down (or silence) the respective gene in living cells, which means to inhibit its expression so that the protein it usually produces is lacking. The effects caused by this lack of protein can then be studied by observation of the treated cells. A widely used strategy for systematic gene silencing is based on the cellular mechanism of RNA interference (RNAi). In essence, RNAi is a cellular protection mechanism to defend the cell against foreign genetic material, e.g., from viruses. This mechanism is induced by free, double-stranded RNA (dsRNA), which normally does not occur in healthy cells. To destroy occurring dsRNAs the cell activates mechanisms to cleave and cut the RNA into pieces, so called small interfering RNAs (siRNAs).

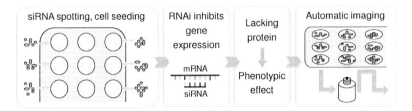

Figure 1.4: Basic steps of an automated siRNA knockdown screen. After spotting the siRNA reagents onto a cover slide, cells are seeded on top. The cells take up the siRNA reagents and RNAi takes place, where in each spot one particular gene is knocked down. As soon as the phenotypic effects of the gene knockdown become visible, automatic image acquisition is started.

Such small single-stranded RNA parts can then bind to other free, single-stranded RNA parts having a complementary base sequence. As soon as an siRNA binds to a matching free RNA fragment, the resulting double-stranded RNA again is cleaved and cut. This mechanism can be utilized to systematically destroy single-stranded messenger RNA (mRNA) which is essential for normal gene expression. For normal gene expression, an mRNA is produced in the cell nucleus by transcription of the DNA region encoding the respective gene. Subsequently, the mRNA is transported outside of the nucleus to the cytoplasm and there binds to a ribosome, which translates the mRNA and produces the respective protein (protein biosynthesis). However, if after mRNA transcription the mRNAs are systematically destroyed by RNA interference, the respective protein can no longer be produced, which is denoted as *post-transcriptional* gene silencing. Thus, to silence a gene based on RNAi, either dsRNAs or siRNAs with a base sequence complementary to the mRNA of the target gene have to be introduced artificially into the cell. Consequently, RNAi gene knockdown experiments require organism-specific libraries providing adequate siRNAs (or dsRNAs) for the selected target genes.

1.2.4 High-throughput high-content screens

To systematically identify genes regulating certain cellular processes (such as, e.g., mitosis), genome-wide siRNA knockdown experiments are performed on living cells. This means, for each single gene in the studied organism (i.e. around 22,000 genes for human cells) at least one knockdown experiment is performed. Such large numbers of experiments can only be managed by automation and optimization of the experimental processes. In essence, an automated live cell image-based siRNA knockdown screen comprises the steps as sketched in Fig. 1.4: first, the gene-specific siRNAs are spotted to cover slides (cell arrays [53] or multiwell plates) by a spotting robot. Such spots typically have a diameter of around 400μm, and each spot (or well) contains the siRNAs required to knock down one specific gene. Next, the cells are seeded over the whole slide, and cells settled on siRNA spots take up the respective siRNAs which initiates RNA interference (see Sect. 1.2.3). Before the actual gene knockdown becomes effective it takes a certain time until the amount of mRNA is sufficiently degraded. As soon as the phenotypes resulting from the knockdown become visible, automated image acquisition is started. To acquire image sequences for all spots simultaneously, the automated microscope consecutively takes an image (or image stack) at each spot, and after the last spot returns to the first spot to start the next loop (see, e.g., [130, 136]). Consequently, the number of spots that can be imaged simultaneously, the time spent at each spot (e.g., to acquire multiple slices), and the desired temporal resolution have to be carefully adjusted to the individual observation requirements. For a more general overview of genome-wide high-throughput screening see, e.g., [26, 52]. Besides genome-wide primary screens (as described above), such automated screening platforms are also applied for follow-up secondary screens (typically studying a subgroup of genes with higher information content).

1.2.5   Analysis of high-throughput images

As described in the previous section, automated microscopy platforms allow imaging of multiple experiments in parallel, and thus, acquiring large amounts of image data relatively quickly. To enable possibly fast analysis of the extensive image data sets acquired in high-throughput high-content screens, automatic image analysis methods can be applied. In the following, we summarize typical image analysis tasks in this context. Usually, in the first step the objects of interest (e.g., cells, cell nuclei, subcellular structures) have to be identified using adequate image segmentation methods. This can also involve the application of preprocessing techniques for deblurring and denoising. The separation of objects and background enables object-based quantification, such as counting the objects or computing object-based measurements (e.g., mean intensity, size, shape). In many applications these analysis steps already provide the main analysis readout: single-object features determined for large populations, possibly computed throughout an image sequence, which serve as a basis for a statistical analysis. However, to study single objects over time or to analyze their movements it is necessary to track the objects, i.e. to determine object correspondences in consecutive time steps. Based on the tracking result the fate of individual objects can be studied. In some applications it is also useful to geometrically align objects before further analysis is performed, which requires methods for image registration. Another typical task is to distinguish object classes on a large scale. To this end, descriptive object features in combination with machine learning techniques can be applied for feature-based classification. Note, however, that developing an accurate analysis approach most often requires the adaptation of existing methods or the development of new methods, since the desired readout as well as the experimental settings differ between applications.

1.3 Main Contributions of this Thesis

In the following we summarize the main contributions of this thesis:

- **Image analysis workflow for cell cycle analysis.** In this thesis we developed an automatic system for quantification of mitotic phase durations as well as nuclear phenotypes, which allows detailed cell cycle analysis. So far, only few approaches have been described for image-based cell cycle analysis on a single-cell basis (see Sect. 2.4.2). In these approaches, on the one hand, a relatively small number of cell cycle phases is distinguished (five phases at maximum, e.g., [35, 149]), and on the other hand, no additional phenotype classes are considered. Here, we distinguish seven regular mitotic phases and up to six additional phenotype classes which allows us to study correlations between temporal and morphological phenotypes, or between early and late phenotypes. Moreover, we present a new method for robust and automatic determination of cell cycle phase durations. Finally, all existing approaches for cell cycle analysis consider 2D image sequences (e.g., [35, 141]), while our scheme copes with 2D and 3D image sequences.

- **Robust segmentation of cell nuclei.** Segmentation of cell nuclei is usually addressed by thresholding, edge- or region-based techniques (see Sect. 3.3.1). We here present an extended region-adaptive thresholding scheme which is optimized for fast processing of images with strong contrast variations. For speed and accuracy optimization we performed a systematic evaluation based on real image data. Furthermore, we extended the basic algorithm to enable segmentation of morphologically difficult cell nuclei. The first extension addresses small dim extranuclei which are attached to normal bright nuclei, making the compound object very hard to segment. The second extension is concerned with merging of small nucleus parts and single detached chromosomes to the main nucleus. Thus, our extended approach allows us to accurately segment morphologically highly abnormal nuclei which are generated by experimental treatment such as gene knockdown. To our knowledge, the segmentation of such complex morphological phenotypes has not been addressed before in existing approaches for cell nucleus segmentation. However, accurate segmentation of the abnormal morphologies is essential to allow correct phenotype classification. We carefully validated our extended segmentation on a large set of real images containing normal as well as abnormal cell nuclei.

- **Mitosis detection for cell tracking.** Cell tracking usually requires strategies to handle splitting events since most living cells have the ability to perform cell division. Previous approaches apply, e.g., geometric active contour methods which naturally allow splits of the contour (e.g, [111, 133, 198]). However, in these methods correspondence finding relies on an overlap of mother and daughter cells which often is not given for cell nucleus tracking. Other approaches apply additional mitosis detection steps after establishing one-to-one correspondences (e.g., [35, 143, 165, 191], see Sect. 3.4.1). We present a tracking approach that is based on the second strategy, and in particular, we developed a new, robust mitosis detection criterion. Our new criterion does not only consider geometric- or intensity-based features but combines morphological and topological features of mother and daughter cell nuclei by a likelihood function. For the evaluation of the mitosis likelihood our method uses automatically determined thresholds which are computed from population averages. Thus, only very few parameters have to be selected by the user. We systematically evaluated the mitosis detection performance based on real image sequences and show that our approach is also very robust for morphologically abnormal image sequences from gene knockdown experiments.

- **Improved feature extraction for multi-dimensional images.** A wide range of descriptive features for cell image classification have been described in previous approaches (e.g., [17, 41, 117, 158, 159], see Sect. 3.5.1). However, in most applications features are extracted based on single 2D images. Extracting features from multi-dimensional images, on the other hand, offers the possibility to integrate information from different dimensions into a set of descriptive features. 3D image features have also been used for cell image classification (e.g., [33]). However, in this work we are dealing with a relatively low number of three to five slices, and typically, 3D image features are designed for a higher number of slices. Therefore, we propose strategies to compute features on automatically selected most informative slices, and additionally, on projected images. We evaluated the performance of the different strategies on a large set of real image data. Furthermore, we also make use of the temporal context by establishing dynamic image features. We define dynamic image features as temporal changes of standard feature values. Including such dynamic image features significantly increases the classification performance as our evaluation on real images proves. Moreover, we present extensive experimental results of the comparison of different feature reduction and classification methods confirming that the methods we chose for our analysis are very well suited. We also experimentally evaluated different strategies to deal with unbalanced training data which often cannot be avoided in real applications.

- **Mitotic phase sequence parser.** To detect temporal mitotic phenotypes as mitotic phase prolongations or shortenings, methods are required to automatically measure the phase durations on a large scale. A suitable approach to determine cell cycle phase durations based on automatic image classification must be able to handle errors in the phase sequence resulting from misclassification. In particular, inconsistent phase sequences should be identified and corrected or excluded in order to produce meaningful results. To check the phase sequence consistency previous approaches applied a small number of relatively simple phase progression rules and compared phase durations to manually set thresholds (e.g., [35, 180], see Sect. 2.4.2). However, this method only allows identification of a small number of possible error types, and thus, is not well suited to robustly determine phase durations on a large scale. In this thesis, we developed a phase sequence parser based on a finite state machine that accepts only biologically consistent phase sequences. Inconsistencies are identified, and if possible, they are corrected. This, on the one hand, allows us to guarantee the consistency of the analyzed phase sequences, and on the other hand, to determine the phase durations automatically. We applied our new scheme to a large number of image sequences and show based on a proof-of-principle experiment that it is able to determine phase durations in high agreement with manual evaluation.

- **Extensive experimental validation on real image data.** We applied our automated analysis system to data from four different screens. First, 40 2D images from a pilot screen were analyzed without considering the temporal context. Note that such an analysis is required for large scale evaluation of genome-wide screens for a rough preselection of interesting genes. In the next step, we analyzed four 3D image sequences of a secondary screen assay on mitotic delay phenotypes. Subsequently, we applied our approach to 48 3D image sequences with partly highly abnormal morphologies, resulting from siRNA and small molecule drug treatment. Again, we determined the mitotic phase durations and correlated the identified temporal phenotypes with automatically detected morphological phenotypes. This experiment was designed as a proof-of-principle experiment and confirmed the overall high performance of our approach. Finally, we applied our scheme to analyze a complete siRNA knockdown screen including 951 3D image sequences. In this application, the main focus was on the analysis of nuclear morphology phenotypes over time. This extensive application on experimental data from four different screening experiments shows that the approach developed in this thesis is not specific to one application but can be applied for different types of analysis. Moreover, we demonstrate the direct applicability of our approach for different image types by analyzing images from two different screening microscopy platforms, and images of two different cell lines.

The results of our research have been published in several peer-reviewed conference proceedings [74, 75, 76, 77, 81, 82, 103] and a journal [80], as well as in two book chapters [73, 79]. The developed approach was also used in several related projects [60, 61, 100, 124, 125].

Chapter 2

Existing Approaches

2.1 Introduction

In this chapter, we review and classify existing approaches for automated analysis of cell microscopy images. In the following sections we proceed from relatively general applications to more specific applications. First, automatic approaches for cell microscopy image analysis in general are considered (Sect. 2.2), then we give an overview of approaches addressing high-throughput high-content experiments (Sect. 2.3) and finally, cell cycle analysis in the context of high-content screening, which is the focus of this thesis, is discussed in detail (Sect. 2.4). We conclude this chapter by summarizing and discussing the limitations of previous approaches (Sect. 2.5).

2.2 Analysis of Microscopy Images in Different Application Fields

Computational approaches for automatic analysis of cell microscopy images have been developed since more than 40 years, and still, the analysis tasks studied at that time remained of high interest until today. Identification of single cells in multi-cell images and computation of cellular features (e.g., area, diameter, eccentricity) has already been addressed, e.g., by Belson et al. [13] in the late 1960s. There, cellular features were computed for 6000 cells and feature distributions were determined and compared for different cell populations using statistical method. Also, automatic classification of different cell types has been studied early, e.g., in 1978 Landeweerd et al. [106] described an approach to classify leukocytes into three different classes based on texture features. With the rapid progress of computational performance as well as the development of devices for digital image acquisition, computational methods for microscopy image analysis became more and more important. The development of automated microscopy facilities in recent years increased the demand for automatic image analysis solutions even further. Today, image analysis methods are applied in various biomedical application fields reaching from basic research in biology to clinical practice. Thereby, the studied structures of interest exhibit different sizes and characteristics, and consequently, require applying different magnification factors and imaging devices, e.g., brightfield, fluorescence, or phase contrast microscopy to study tissue sections, cell populations, single cells, or subcellular structures and particles (e.g., vesicles or virus particles). Even single molecules and atoms can be studied using transmission electron microscopy. In this work we focus on applications in the range of multiple cells to subcellular structures.

An increasingly important clinical application field is histo- and cytopathology. Histopathology considers tissue sections, mostly stained with immunohistochemical stains (Haematoxylin and Eosin), and imaged with classical brightfield microscopy or tissue slide scanners. Typical analysis tasks are segmentation of tissue substructures, such as, e.g., cell nuclei, stroma, epithelial tissue, and glands (e.g., Datar *et al.* [45]), and classification of tissue types, for example, to identify pathological tissue (e.g., Chebira *et al.* [29], Ficsor *et al.* [55], Glotsos *et al.* [62]). Tissue sections have also been used in high-throughput experiments based on fluorescence microscopy, e.g., to establish automatic tissue scanning systems for routine diagnostic biopsies (see Nattkemper *et al.* [134, 135]) or to study epithelial differentiation (see Grabe *et al.* [68]). In cytopathology, cells extracted from body fluids, smears, or needle aspirates, are examined for clinical diagnosis. In this application, image analysis can be used, e.g., to classify different cell types (e.g., Würflinger *et al.* [189]).

Images of subcellular structures are analyzed, e.g., to determine the localization of proteins within a cell which provides important information on protein function (*location proteomics*). In the past ten years, different approaches for automatic classification of subcellular structures have been published, in particular, by Murphy *et al.* [16, 17, 33, 89, 90, 132, 174]. Work has been done, e.g., on determining optimal feature sets for classifying subcellular structures [16, 17, 33, 90, 132, 174] and on different types of classification methods, such as neural networks, support vector machines, and ensemble methods [89, 90]. Their results have been of use for various other microscopy image analysis approaches, e.g., for analysis of high-throughput experiments. Classification of subcellular structures for location proteomics has also been studied, e.g., by Danckaert *et al.* [43], Conrad *et al.* [41], and Hamilton *et al.* [69].

In many fields of basic research in biology large-scale automated experiments are performed providing huge sets of multi-cell images. There, usually the goal is to compute measurements on a single-cell basis and to analyze those measurements for a high number of cells. For studying various biological questions, different treatments and imaging techniques are applied. Thereby, either static images of fixed cells or dynamic images of living cells are acquired. Below, we give an overview of existing approaches for analyzing multi-cell images in the context of high-throughput, high-content screens (Sect. 2.3). In a separate section (Sect. 2.4) we describe cell cycle analysis which can be considered as a sub-field of multi-cell image analysis and is the focus of this thesis.

2.3 Analysis of High-Throughput Multi-Cell Images

In recent years, a large number of approaches to automatically analyze multi-cell images acquired in high-throughput high-content screens have been published. Since the application fields and analysis tasks are quite different, many different methods have been applied. Approaches for multi-cell image analysis can generally be classified into two groups: first, approaches that do not take into account the temporal context for image analysis and focus on the extraction of features and classification of objects based on static images, and second, approaches which exploit the temporal context based on cell tracking and trajectory analysis. Thus, the first group of approaches is typically used for population-based analysis, while the second group of approaches enables single cell-based analysis over time. In the following subsections we discuss both groups of approaches in more detail.

2.3.1 Analysis without exploiting the temporal context

In this group of approaches the analysis can be subdivided into the steps of (1) cell segmentation, (2) feature extraction, and (3) feature mining in terms of statistical analysis of the features or object classification. One typical goal is to study the effects of experimental treatment on a cell population by comparing population-based features of treated and non-treated groups using, for example, RNAi knockdown (e.g., Zhou *et al.* [201], Goshima *et al.* [67], Matula *et al.* [123, 124], Wang *et al.* [178]), protein overexpression (e.g., Harada *et al.* [71]), or drug profiling (e.g., Lindblad *et al.* [117], Perlman *et al.* [145, 146], Loo *et al.* [119]). Another typical goal is to distinguish different cell types or tissue types for clinical applications (e.g., Adiga *et al.* [7], Chebira *et al.* [29]). Usually, the labeled structures of interest are, on the one hand, cell nuclei, and on the other hand, additional subcellular structures imaged in separate channels such as, e.g., the cytoplasm or cytoskeleton (e.g., Lindblad *et al.* [117], Jones *et al.* [94], Wang *et al.* [178], Szafran *et al.* [167]), the cell membrane (e.g., Han *et al.* [70]), the spindle apparatus (e.g., Goshima *et al.* [67]), or proteins indicating a specific cellular response (e.g., virus infection, Matula *et al.* [123, 124]). Since different subcellular structures usually have differing morphological properties, the segmentation in different imaging channels often requires specific segmentation algorithms such as, e.g., threshold- or edge-based segmentation for cell nuclei and region- or deformable model-based segmentation for complete cells. Thereby, segmentation of the cytoplasm or the cell boundary is usually more challenging than cell nucleus segmentation (e.g., [70, 94, 117, 123, 124, 178]). More details on different approaches for cell segmentation are given in Sect. 3.3.1 below.

After segmentation, in the next step cellular features are computed based on the identified single objects. The extracted features include characteristics of the objects such as size, shape, or intensity distributions (e.g., texture features). In the first group of approaches, segmentation and feature extraction is based solely on the nucleus channel in 2D (e.g., Wheeler *et al.* [184], Neumann *et al.* [136], Zhou *et al.* [199], Gambe *et al.* [58], Walter *et al.* [177], Kim *et al.* [99]) or 3D (e.g., Roula *et al.* [160], Lin *et al.* [116]) images. The extracted nucleus features are analyzed by directly comparing feature distributions between different experimental groups, e.g., to distinguish normal and malignant nuclei using T-test statistics [160] or to detect hits in loss-of-function screens [184]. Alternatively, machine learning techniques are applied to classify cell nuclei based on the extracted features, e.g., into different cell types [116] or different cell cycle phases [58, 99, 136, 177, 199] (see Sect. 2.4). The second group of approaches not only consider the nucleus channel but also use image features from additional channels. For the additional channels, the most frequently extracted features are the fluorescent signal intensity in the segmented regions [71, 124, 167], location and intensity related features of spot-like structures (e.g., spliceosomes, centrosomes, fluorescent *in situ* hybridization signals) [7, 145, 146], or more specific shape and intensity features depending on the structure of interest (e.g., the spindle [67]). Based on the extracted features a statistical analysis of the feature distributions between different treatment groups is performed. A feature set from different channels has been used, e.g., to quantify the separability of experimental treatment groups based on classifiers in order to create cellular drug response profiles [119] or to determine optimal feature sets [70]. Multi-channel features have also been applied for directly classifying single cells into different phenotype classes. For example, in [117, 178, 201], cells were classified automatically into different stages of the activation of a fluorescently labeled cellular signaling protein using supervised learning methods.

In this context, image analysis software platforms have been developed which provide tools for large-scale analysis. The goal is to provide easy to use image analysis tools that are suitable for a possibly wide range of applications and can be combined into individual analysis workflows. Such software platforms particularly intend to allow users that have little knowledge and experience in the field of image processing to analyze large-scale experiments on their own. Therefore, a graphical user interface as well as tools for visualization and export of the results are usually included. A well known open source platform is CellProfiler [2, 25, 95, 105]. This software provides applications for image preprocessing, segmentation, feature extraction, classification, statistics, and visualization. Recently, Liron *et al.* [118] started another open source project with comparable goals. Also the public domain image processing software ImageJ [156] is often used to perform basic image analysis tasks, however, it is not adapted for high-throughput analysis. Furthermore, commercial software platforms are offered, e.g, by microscope companies or independent software companies providing solutions for frequently occurring analysis tasks. However, note that for more specific analysis tasks general software platforms often do not provide an appropriate solution.

2.3.2   Temporal analysis using tracking

The approaches described in this section deal with the analysis of temporally resolved image sequences from live cell microscopy. Importantly, all these approaches involve object tracking, i.e. determining correspondences between cells in consecutive image frames and thus establishing the motion path (or trajectory) of a cell. The determined trajectories, on the one hand, can be used to analyze cellular features over time on a single-cell basis, and on the other hand, provide information on the properties of cellular movement and replication (i.e. cell motility and proliferation). Tracking is commonly performed either based on a deterministic two-step approach which consisits of cell segmentation followed by correspondence finding (e.g., Sigal *et al.* [165], Chen *et al.* [35], Yan *et al.* [190], Gordon *et al.* [65], Chen *et al.* [36]), or by evolution of deformable models (e.g., Zimmer *et al.* [202, 203], Mukherjee *et al.* [131], Debeir *et al.* [47], Dufour *et al.* [50], Yang *et al.* [191], Padfield *et al.* [141], Bunyak *et al.* [22]) or probabilistic models (e.g., Li *et al.* [111], Wang *et al.* [182]). A detailed review on different approaches for cell tracking is given in Sect. 3.4.1 below.

A large group of approaches deal with the analysis of live cell image sequences acquired with *phase contrast microscopy*. There, the analysis task often involves a long-term study of cellular motility and proliferation behavior. Due to long observation times, in these images the cell density often is high, and cells are touching and overlapping which makes the analysis particularly challenging. Many approaches apply deformable or probabilistic model-based tracking methods to tackle these problems. The effects of experimental treatment on cell motility have been studied, e.g., for cells exposed to different concentrations of a growth factor in Li *et al.* [110, 113] or an *anti-motility* drug in Debeir *et al.* [47]. Other studies focused on movement analysis of epithelial cells during *in vivo* wound healing (e.g., Bunyak *et al.* [22]) or cells performing autophagy (*self-cannibalism*, e.g., Wang *et al.* [182]). Another important application field is stem cell research where tracking is applied to create cellular lineage maps which provide important information such as the symmetry of a division map or the division time (e.g., Li *et al.* [111, 112]). Motility analysis was performed by comparing the mean displacements (mean velocities) [22, 113, 182] or by statistical analysis of other features characterizing the movement patterns (e.g., the convex hull of a trajectory or the maximum distance to the origin) [47]. In Li *et al.* [111] a tracking approach based on different motion models was developed which allowed classifying the cellular movements into different motion types. Studying cell proliferation naturally requires the reliable detection of mitosis events. In phase contrast microscopy images, mitosis events usually can be identified based on intensity and shape features since there, mitotic cells typically appear as bright, round objects (e.g., Gallardo *et al.* [57], Yang *et al.* [191], Li *et al.* [112]). Combined approaches for tracking and mitosis detection in phase contrast microscopy images have been described, e.g., in [57, 191]. In Gallardo *et al.* [57] tracking was performed based on the overlap of cell regions, and hidden Markov models were used to classify cells into normal cells, dead cells, and mitotic cells. In Yang *et al.* [191] geometric active contours for tracking and a supervised classifier for mitosis detection were combined.

Images acquired by *fluorescence microscopy* often provide a higher information content compared to phase contrast microscopy, since multiple cellular structures can be labeled with different fluorescent dyes and imaged simultaneously in different channels. Consequently, time resolved analysis of multi-channel fluorescence microscopy images often is targeted on analyzing cellular features from different channels over time. For example, in Sigal *et al.* [165] the nuclear accumulation rate (nuclear fluorescence/nuclear shape) was compared for different channels, and in Gordon *et al.* [65] the maturation rate was determined based on total fluorescence.

In Chen *et al.* [36] interactions between different types of cells imaged in different channels were quantified. For fluorescence microscopy images, cell tracking is often performed based on the nucleus channel as cell nuclei usually have a high contrast and are well separable. If the displacements of cell nuclei are small w.r.t. the nucleus size, then relatively simple criteria are sufficient for correspondence finding such as, e.g, the Euclidean distance [165] or the area overlap [65]. However, if displacements are larger, then additional criteria are required to determine correspondences, e.g., in Chen *et al.* [36] correspondences were determined based on the differences of feature vectors including volume, radius and centroid location of the tracked cells. Also, if the signal-to-noise ratio in the nucleus channel is low, additional preprocessing steps are required to improve the detection of cells, e.g., in Padfield *et al.* [142] a wavelet-based denoising approach was used.

2.4 Cell Cycle Analysis

One particular group of high-throughput studies aim to elucidate the mechanisms cell cycle regulation. As described in Sect. 1.2.1 above, the cell cycle can be generally split into interphase and mitosis (M-phase). Interphase is subdivided further into G1, S, and G2 phase, and mitosis usually is split up into six mitotic phases. Since cell cycle progression is related to serious diseases (e.g., cancer), research in this field is of high common interest. As described in the previous section, mitosis detection is an important part of several automatic analysis approaches. But also more detailed cell cycle studies considering multiple subphases of mitosis or interphase have been analyzed using automatic image analysis approaches. Existing approaches for cell cycle analysis can be split into two groups: approaches which do not take into account the temporal context for image analysis, and approaches that exploit the temporal context. Both groups will be described below.

2.4.1 Analysis of cell cycle phases without temporal context

Determining cell cycle phases based on image information usually requires image feature-based classification. Thus, the general workflow for cell cycle phase analysis typically consists of three steps: (1) segmentation, (2) feature extraction, and (3) classification. The approaches described in this section are all concerned with the classification of mitotic subphases, i.e. prophase, prometaphase, metaphase, anaphase, and telophase. To distinguish these subphases, classification is typically performed based on the nucleus channel. However, in Zhou *et al.* [199] information from three channels was used for classification based on logical rules. In the three

channels the following structures were visualized: (1) the DNA providing information on the chromosome location, (2) actin (part of the cytoskeleton) to visualize the
cytoplasm, and (3) microtubules, which are evenly distributed in interphase and form
the mitotic spindle during mitosis. Segmentation was performed in all channels using
a Laplacian of Gaussian (LoG) operator for edge detection. Then, relatively simple
binary features (object visibility for each channel) and shape features (eccentricity)
were used to classify cells into three classes (interphase/prophase, metaphase, and
anaphase/telophase). In Tao *et al.* [168] four imaging channels were used, visualizing
(1) the DNA, (2) a histone which allows distinguishing interphase and prophase, (3)
an indicator for DNA synthesis, and (4) tubulin, i.e. microtubules and mitotic spindle. Segmentation was performed using watershed transform. Shape and intensity
features were extracted from all channels and, after feature selection, classification
was performed based on the best 59 features. A support vector machine (SVM)
classifier was applied to classify the cells into prophase, prometaphase, metaphase,
anaphase/telophase, and a phenotypic class representing abnormal shapes.

Other approaches rely solely on the nucleus/DNA channel, which demonstrates
that the DNA channel in principle is sufficient to distinguish mitotic subphases. To
enable a subdivision into detailed mitotic phases the spatio-temporal resolution of
the images naturally plays an important role: the higher the resolution, the more
subphases can be distinguished. For example, only if the image resolution is high
enough to resolve the beginning chromosome condensation in prophase it is possible
to distinguish interphase and prophase. In Neumann *et al.* [136] cell nuclei were
segmented using automatic adaptive thresholding, and classification was performed
into five classes (interphase, mitosis, apoptosis, clustered nuclei, and artefacts) based
on a feature set of about 200 texture and shape features. For classification SVMs
were applied and the number of samples per class was studied over time. In Kim *et
al.* [99] tophat filtering was used for segmentation and cell nuclei were classified into
four classes (interphase, metaphase, telophase, and apoptosis). There, 62 shape- and
intensity-based features were extracted and classification was performed using logis-
tic regression. A more detailed classification into mitotic subphases was achieved
in Gambe *et al.* [58], where eight classes were distinguished (interphase, prophase,
prometaphase, metaphase, early anaphase, anaphase, telophase, and cytokinesis)
based on 26 features using multivariate linear discriminant analysis. However, in
[58] a threshold-based segmentation was used, and the thresholds were selected manually for the around 100 processed images, in comparison to the above described
approaches which used automatic segmentation approaches. Consequently, for large-
scale analysis the approach of Gambe *et al.* [58] would have to be extended by an

automatic segmentation method.

2.4.2 Analysis of cell cycle phases by exploiting the temporal context

In this section we describe approaches which exploit the temporal context for the analysis of cell cycle phases using tracking. Partly, this type of analysis has already been addressed in Sect. 2.3.2 above, where mitotic cells were tracked and classified to study proliferation rates (e.g., [57, 191]). However, there, mitosis was treated as one class and no further subphases were distinguished. The approaches discussed in the previous Sect. 2.4.1 distinguished several cell cycle phases but, on the other hand, temporal information was not provided or not considered for the image analysis. In this section we describe approaches that combine both attempts to enable single cell-based analysis of cell cycle progression. This allows studying correlations of phenotypes with cell cycle phases, or the effects of experimental treatment on cell cycle phase progression (e.g., to study drugs which inhibit the replication of cancer cells [143]). Note that here we describe approaches that determine subphases of interphase as well as approaches for classification of mitotic subphases.

An approach to split up the cell cycle into the interphase subphases (G1-, S-, G2-phase), and M-phase has been described by Padfield *et al.* [141, 143, 144]. There, a dynamic cell cycle phase marker was used to fluorescently label the cells instead of visualizing the DNA as done in most other approaches. This marker binds to a protein which moves between the nucleus and the cytoplasm during different cell cycle phases and allows distinguishing G1-, S-, and G2-phase based on the intensity relation between nucleus and cytoplasm. Cell nuclei in G2-phase were segmented and tracked using 3D level set active contours, where the 2D image sequences were considered as spatio-temporal volume data. Subsequently, cell nuclei in all other phases were segmented and linked to the segmentations of the nuclei in G2-phase to establish complete cell lineage trees. In [143] the segmentation of the other phases was done in two steps: first, nuclei in M- and G1-phase were segmented using region growing, as in both phases cell nuclei have a relatively high contrast. Nuclei in S-phase, on the other hand, have the same intensity as the cytoplasm, and therefore, the positions of the nuclei had to be determined by interpolation from adjacent images. In the second step, the segmented tree parts were linked based on minimum Euclidean distance. However, this linking strategy is prone to association errors if neighboring cells are very close or partly overlap. Consequently, Padfield *et al.* extended their approach by replacing the Euclidean distance-based phase linking with an intensity-based model for phase linking [141, 144]. This model represents the smooth transition of bright cell nuclei in M-phase to gradually dark cell nuclei in S- and G2-phase based on a sigmoid function. The model is used as a cost function and the phase linking result is finally determined by searching the paths with the lowest costs between G2 segmentations.

For the classification of cell nuclei into mitotic phases several approaches have been described by Wong *et al.* [35, 149, 180, 181, 190, 192, 200]. There, generally four classes were distinguished, namely interphase, prophase, metaphase, and ana-/telophase, and in a few approaches a fifth class "arrested metaphase" was considered [148, 149]. All approaches involve segmentation, tracking, feature extraction, and classification based on machine learning. Chen *et al.* [35] presented an approach with watershed-based segmentation and tracking using the Euclidean distance as well as size and position constraints. A set of 40 image features, including intensity-, shape-, and texture-based features were extracted and reduced to eight features using principal component analysis (PCA). Classification was performed using a k-nearest neighbor (KNN) classifier. Finally, three heuristic rules were applied to correct classification errors by checking the biologically defined sequence and duration of phases. Thereby, arrested metaphases could be identified for metaphase sequences longer than a predefined maximum duration. An approach with an improved overall classification performance was proposed by Yang *et al.* [192], where the KNN classifier was replaced by an artifical neural network (ANN) classifier. Tracking was performed based on the mean shift algorithm and a Kalman filter. The feature set was also extended by steerable filter features which can be computed for multiple arbitrary directions with relatively low computational effort. In total, 44 features were extracted and reduced to 18 features using PCA. Yang *et al.* showed that the overall performance of the classifier could be improved using the extended feature set and the ANN classifier. Pham *et al.* [148, 149] studied the application of Gaussian mixture models and Markov models based on seven simple features (e.g., mean intensity, perimeter, compactness). There, it was shown that these classifiers are superior to KNN and k-means clustering. Hidden Markov models (HMMs) were also applied by Yan *et al.* [190] and Zhou *et al.* [200], and compared to a maximum likelihood classifier and a KNN classifier. Again, the performance of the HMM classifier turned out to be best. An extension of HMM classification in the context of cell cycle analysis was presented by Wang *et al.* [180] who proposed to deduce the current state of the model not only based on the previous state (as it is the case for normal HMMs) but to include also the subsequent state. Therefore, the classification scheme presented in [180] was denoted as context-based mixture model (CBMM). Comparing the CBMM with traditional classifiers (KNN, ANN,

SVMs), Wang *et al.* showed that the CBMM outperformed the other classifiers, in particular, regarding the classification of prophase. Recently, Wang *et al.* [181] proposed an approach using updatable, so called, online support vector machines (online SVMs). This classifier has the advantage that new training data can be used to update the previously trained classifier instead of requiring a complete retraining. Thus, if a missclassified sample is observed, the classification can be manually corrected and the corrected sample is fed back to the classifier for a training update. Also, weighted SVMs were used in [181], i.e. the trained classes are assigned different weights according to their sample number in the training set which can improve the performance for unbalanced training sets. Thus, the influence of the usually very frequent interphase samples on the final trained classifier can be reduced by assigning a lower weight in favor of rare classes such as, e.g., meta- or anaphase, for which higher weights can be assigned. A comparison with normal SVM and non-weighted online SVM classifiers showed that the weighted online SVM classifier yielded the highest accuracies.

2.5 Summary and Conclusion

In this chapter, we presented an overview of existing approaches in the field of cell microscopy image analysis with a focus on cell cycle analysis. We classified the approaches into groups with regard to application and workflow, starting with a broad range of applications in different biological areas (Sect. 2.2), then focusing on the context of large scale high-content experiments (Sect. 2.3), and finally going into detail on approaches for cell cycle analysis (Sect. 2.4). Particularly, we distinguished approaches that exploit the temporal context and approaches that do not. Both types of approaches typically include methods for cell segmentation, image feature extraction, and in some cases also classification. Approaches that exploit the temporal context additionally include methods for cell tracking with mitosis detection. Our review shows that in the area of cell cycle analysis so far only few approaches have been described for temporal analysis of image sequences. Also, only up to five mitotic phases were distinguished in such approaches. In addition, most approaches did not consider morphological phenotypes occurring as a result of experimental treatment. In Tao *et al.* [168] one class representing abnormally shaped nuclei was introduced. However, classes such as, e.g., different abnormal cell cycle phases, cell death, or other nuclear morphology phenotypes were not studied. Also, note that the approach of Tao *et al.* [168] is based on static images and does not consider the temporal context. Moreover, in previous approaches phase sequences were determined, but not further analyzed in detail, e.g., to determine temporal phenotypes or irregularities of mitotic progression. In Wong *et al.* [35, 180, 190, 200] three phase progression rules were introduced to check for obvious violations of the phase sequence consistency. For metaphase the phase durations were determined and compared to a manually set threshold. However, this approach is not suited to robustly determine phase durations on a large scale since such rules only capture a small range of possible errors. In addition, the phase durations should be statistically compared between treatment groups rather than comparing them to a manually set threshold, to make the system independent of previous assumptions on the treatment effect.

Furthermore, most approaches for high-throughput analysis are based on 2D images or image sequences, and, in particular all approaches for cell cycle analysis are concerned with 2D image sequences. To enable the analysis of 3D image sequences, e.g., from a confocal fluorescence microscope, extensions are required. For cell nucleus segmentation all previous approaches deal with normal, morphologically non-disturbed nuclei. However, experimental treatment such as gene knockdown can produce highly abnormal morphologies posing new challenges for segmentation algorithms. Tracking with mitosis detection has been addressed in several previous approaches, where either methods were applied which naturally allow splits (e.g., [22, 50, 111, 113]), or mitosis was detected based on morphological criteria (e.g., [35, 165]) or based on supervised classification (e.g., [57, 112, 191]). However, in previous approaches mitosis usually was assumed to be a smooth transition of one object into two objects, which is not the case for cell nuclei in images with a relatively low temporal resolution. In this case, the main problem is to distinguish nuclei appearing because of a mitosis from nuclei entering from the image border. Therefore, characteristic features of mitotic cell nuclei need to be taken into account, such as, e.g., the size, intensity, and topology of potential mother and daughter nucleus. Approaches for classification of cells in image sequences often only consider object features extracted from the current image frame. An exception are approaches using HMMs for classification since there, information from the previous image frame is naturally considered. In Wang *et al.* [180] information from previous and subsequent image frames was considered using a context-based mixture model. However, HMM-based approaches have the drawback that they are trained based on phase sequences and for a complex model many training sequences are required. Moreover, HMMs are less suited to determine phase durations that strongly differ from the typical phase durations in the training data since the learned phase transition probabilities are adapted to the phase durations in the training set.

In conclusion, a wide range of approaches for cell microscopy image analysis have been described within the past four decades. However, there are still a number of open issues and limitations, in particular, regarding the detailed temporal analysis of cell cycle phases. In particular, the classification of morphological phenotypes occurring in high-throughput knockdown experiments has not been extensively studied.

Chapter 3

Methods

3.1 Introduction

In this chapter, we describe our approach for automatic analysis of cell cycle phases and cell cycle phase durations based on 2D and 3D fluorescence microscopy image sequences. Since we are concerned with a relatively complex analysis task, that is, the determination of detailed cell cycle phases and their durations, our approach is based on a number of different methods from the areas of computer vision and pattern recognition which we combined in a complex analysis workflow. Thereby, methods at the beginning of the analysis workflow are closely associated to the images and work directly on the image data, while towards the end of the workflow methods become more and more abstracted from the actual image data and rather process the extracted information. In the following, we first present the general workflow, where we distinguish between the analysis of 2D and 3D image sequences (Sect. 3.2). Then, we successively address the single image analysis steps, beginning with segmentation (Sect. 3.3) and tracking (Sect. 3.4) of cell nuclei. Next we describe the extracted image features, discuss feature normalization and reduction (Sect. 3.5), and provide an overview of the used classification method (Sect. 3.6). Subsequently, we present our approach for phase sequence processing based on a finite state machine (Sect. 3.7). Finally, our overall strategy for data handling and data visualization is described (Sect. 3.8) and we conclude the chapter with a brief summary.

3.2 General Workflow

In this thesis, we consider two image analysis tasks resulting from biological screening experiments. The first task is the analysis of 2D images with respect to the distribution of different classes of cell nuclei without considering temporal information. In our case, we distinguished four different classes (interphase, mitosis, apoptosis, and multi-nucleated), and we have developed a straightforward image analysis workflow consisting of segmentation, feature extraction, and classification. The second task is to determine cell cycle phase durations for seven cell cycle phases and morphological phenotypes based on 3D image sequences. Here, a much more complex workflow had to be developed, which includes cell nucleus tracking and phase sequence analysis. Both workflows are presented in detail in the next two sections. More details on the biological questions motivating our approach are given in Chapt. 4.

3.2.1   Analysis of 2D images

For classification of cell nuclei into different phenotype classes based on 2D multi-cell images we apply the general workflow shown in Fig. 3.1. First, segmentation is performed to identify and label single cells. Based on the resulting 2D regions-of-interest, image features are extracted, which provide a general numeric description for each cell. The extracted features are then used as input for automatic classification methods. In our case we use supervised learning, which means that the classifier initially has to be trained with a set of manually labeled examples, and subsequently it can be applied to automatically classify unclassified samples. As a result we determine for each cell nucleus in each input image a phenotype class (e.g., interphase, mitosis). For experimental results using this type of analysis see Sect. 4.2 below. If image sequences are analyzed and the temporal information is of interest, an additional step for cell tracking is required, as it is the case in the workflow described in the following section.

3.2.2   Analysis of 3D images

The main goal of this thesis is to enable automatic determination of cell cycle phase durations based on 3D multi-cell image sequences (see Sect. 4.3 and 4.4 below). To this end, we developed the workflow shown in Fig. 3.2. Here, the input image sequences consist of three to five slices per time step, however, all image analysis steps are performed in 2D for the following reasons. First, 2D image analysis methods are generally faster than 3D methods and since our goal is high-throughput analysis, processing speed is important. Second, the relatively low number of slices (three to five), as well as the high unisotropy of resolution (the resolution of 1024×1024 in x-y-direction is much higher than in z-direction) argue against the application of

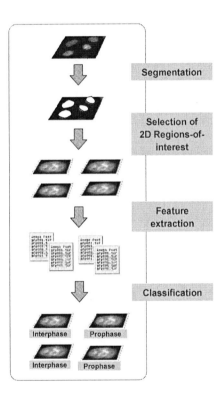

Figure 3.1: General workflow for cell nucleus classification based on 2D image sequences. Here, the temporal context is not considered.

3D image analysis methods. The number of acquired slices per time step was not chosen higher because of restrictions of the biological experiments. Acquiring an increased number of slices requires to illuminate the fluorescently labeled cells for a longer period. However, in live cell fluorescence microscopy the duration of illumination must be minimized to prevent the cells from being damaged by the laser light (*phototoxicity*, see, e.g., [130]). Moreover, using an automatic microscopy platform, an increase of the number of slices acquired at each location means a decrease of the number of simultaneously observed locations, or a decrease of the temporal resolution (see Sect. 1.2.4). On the other hand, using a confocal microscope necessitates the acquisition of multiple slices since the structure of interest, i.e. the chromatin, moves in z-direction throughout the cell cycle, and thus, also moves out of focus. Cells growing on culture plates are flat in interphase and the chromatin is located close to the plate. However, when cells start to divide, they round up and the chromatin moves upwards, i.e. away from the culture plate. Consequently, the chromatin in a mitotic cell cannot be imaged at the same focal plane as the chromatin in an interphase cell (see Fig. 3.3). Since cells in cell cultures usually are not synchronized, and thus, can be in different cell cycle phases at a certain time point, we acquire three slices to ensure that for each cell the chromatin structure is captured at least in one slice. This slice usually is sufficient to determine the cell cycle phase of the respective cell. However, this slice, providing the most information for phase determination (i.e. the most informative slice), has to be determined for each cell in each image frame separately.

In the first step of our workflow a maximum intensity projection (MIP) of the 3D images is performed at each time step, resulting in 2D image sequences (Fig. 3.2). For MIP, at each pixel position the maximum intensity is selected from the different slices. Next, segmentation and tracking are performed on the MIP images. Performing segmentation on the MIP images ensures that for each slice the structures of interest lie inside the segmented region, and consequently, the segmented region can be used as a region-of-interest in all slices. Based on the segmentation and tracking result, 3D regions-of-interest (ROIs) are selected in the original 3D image sequences for all cell nuclei. Within these 3D ROIs the slice which contains the most information is selected. These most informative slices are then used for extraction of image features. Alternatively, also the MIP images can be used for feature extraction. However, for the extraction of texture features the most informative slices are more suitable since MIP generally leads to a blurring of textural structures which implies a loss of information. In this thesis, we use both strategies: feature extraction based only on the most informative slices (see Sect. 4.5 and 4.3) and feature extraction

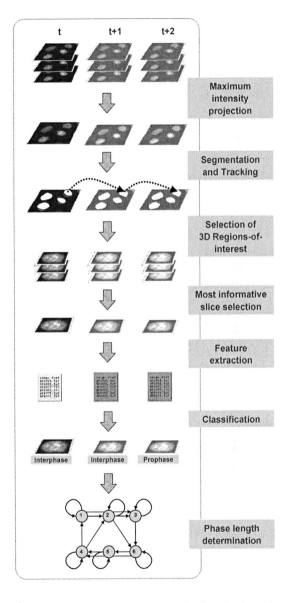

Figure 3.2: General workflow for determination of cell cycle phase durations based on 3D image sequences. Here, the temporal context is exploited.

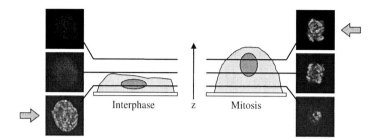

Figure 3.3: Live cell imaging of the chromatin structure using a confocal fluorescence microscope. In interphase, cells are flat and the chromatin (dark gray disk) can be imaged close to the culture plate. During mitosis the cell rounds up and the chromatin moves away from the culture plate in z-direction. Thus, multiple slices (focal planes) have to be acquired to have the chromatin in focus at least in one slice. The gray arrows indicate the most informative slice for each cell.

based on the most informative slices as well as on the MIP images (see Sect. 4.4). Besides commonly used static image features we also extract and use dynamic image features.

After feature extraction all cell nuclei are classified into different cell cycle phases. Prior to classification, for each feature the feature values are normalized to a mean value of zero and unit variance. Combining the tracking and classification results provides a sequence of cell cycle phases for each cell nucleus. For cells performing mitosis within the observed time period, the phase sequences can be represented as a tree-like structure (cell lineage trees or cell pedigrees). To determine the durations of the consecutive phases, the phase sequences are processed using a finite state machine which constitutes a *phase sequence parser*. In addition, the parser detects and corrects classification errors that cause inconsistencies of the phase sequence (i.e. deviations from the biologically valid sequence). Finally, a statistical analysis of the phase length distributions is performed to detect image sequences with significant phase prolongations or shortenings.

3.3   Cell Segmentation

3.3.1   Introduction

Cell segmentation is a fundamental task for automatic analysis of cell microscopy images. In recent years, a large variety of approaches for segmentation of the cell nucleus and the cytoplasmic region (or the whole cell) in images from different types of microscopes have been described. Cell nucleus segmentation is usually less difficult since the shape of nuclei is compact, approximately round or elliptic, and nuclei are typically visualized with a relatively high contrast. Segmentation of the cytoplasm, on the other hand, is generally more challenging because the shape of the cytoplasm is more complex and it often exhibits a lower contrast (depending on the used microscope and staining). A common problem of cell segmentation is the separation of cell clusters. In the following, we first describe main methods for cell segmentation as used in previous work, and second, provide an overview of our segmentation approach.

Existing approaches for cell segmentation

Many approaches apply *thresholding techniques* (see also Sect. 3.3.2 below) for initial segmentation, particularly, of cell nuclei. Either one global threshold is determined and applied to the whole image, or adaptive thresholds are automatically computed for local image regions which can significantly improve the segmentation result (e.g., in case of uneven illumination or staining). Also *edge-based* segmentation approaches have been applied for cell segmentation. For edge detection usually the magnitude and direction of the local image gradient are computed using convolution kernels to determine partial derivatives of an image. For cell and cell nucleus segmentation common edge detection operators such as the Canny operator and the Laplacian-of-Gaussian (LoG) operator have been applied, for example, in Perlman *et al.* [146] and Li *et al.* [111].

Region-based segmentation methods constitute another group of techniques commonly used for cell segmentation. Such approaches determine spatially connected regions of pixels with similar properties. One approach for initial pre-segmentation is the top-hat transform (rolling-ball filter). Rolling-ball filtering simulates a ball rolling beneath the intensity profile, thereby removing all peaks the ball cannot touch. The filtered image is subtracted from the original image providing a rough segmentation of bright objects smaller than the ball diameter (which is equivalent to the top-hat transform). This technique has been applied, e.g., for initial cell segmentation in phase contrast images (Li *et al.* [111]) and for contrast enhancement of fluorescently stained cell nuclei (Zhou *et al.* [200, 201]). Conventional *region-growing* is a conditional dilation of a given seed region, where neighboring pixels are added to the region if they fulfill certain similarity constraints (e.g., similar intensity value). Region growing has been applied, for example, for segmentation of the cytoplasm or complete cells in fluorescence microscopy images (e.g., Lindblad *et al.* [117], Matula *et al.* [123, 124], Fenistein *et al.* [54]), and for segmentation of nucleus compartments in cytopathological images (e.g., Wu *et al.* [187]).

A popular region-growing method is the so called *watershed transform* [175] which is often used for cell image segmentation. In comparison to conventional region growing methods, the watershed transform determines regions based on intensity levels instead of neighborhood levels. The intensity surface of an image is interpreted as a landscape and water floods the surface starting at intensity minima (catchment basins). As soon as the water fronts of two neighboring catchment basins meet, a crest is built between the regions forming the watershed lines which provide the segmentation result. Since this method usually results in oversegmentation, a seeded or marker-based watershed transform can be used. With this approach the flooding process starts at certain seed points which are artificially introduced minima. Seeded watershed transform has been applied, e.g., for 3D cell nucleus segmentation based on the gradient filtered image (Wählby *et al.* [176]) or for cytoplasm segmentation based on the inverted image where prior segmented cell nuclei are used as seed points (e.g., [117, 123, 124, 201]). The watershed transform has also been applied for the separation of clustered cell nuclei after initial pre-segmentation in 2D (e.g., Malpica *et al.* [121], Lindblad *et al.* [117], Li *et al.* [109], Zhou *et al.* [200]) and 3D images (e.g., Adiga *et al.* [6, 7], Lin *et al.* [115]). Depending on the image properties the watershed transform is applied to the original intensities, the gradient filtered image, or the distance transform of a binary image. In the latter case, typically an Euclidean distance transform is used which assigns each foreground pixel the distance to the closest background pixel as new intensity value. This allows separating close and touching cells as long as they are shaped approximately round. Also, combined strategies such as the gradient-weighted distance transform have been applied [115]. Separation of clustered nuclei and merging of oversegmented nuclei has furthermore been addressed using geometric properties of the segmented regions (e.g., size, shape), and split-merge decisions have been performed, e.g., based on decision rules (Zhou *et al.* [201]) or supervised learning (Lin *et al.* [116], Beaver *et al.* [12], Zhou *et al.* [200]).

Segmentation based on *deformable models* allows integration of a priori knowledge about the object geometry to improve robustness and accuracy. The most popular example are active contour models (e.g., Kass *et al.* [97], McInerney and Terzopoulos [126]) which are frequently used for cell image segmentation. With active contour methods, objects are modeled as closed curves which evolve over time to match the object boundaries as good as possible. To this end, an energy functional has to be minimized which consists of two terms representing an internal energy and an external energy. The *internal* energy (regularization term) defines physical constraints of the contour (e.g., stiffness) which allows controlling the smoothness of the curve. The *external* energy (data attachment term) drives the curve towards features of interest in an image (e.g., edges). Active contour methods, in particular, geometric active countours (Osher and Sethian [138, 161]), have been successfully applied for cell and cell nucleus segmentation in 2D (e.g., Solorzano *et al.* [137], Zimmer *et al.* [202], Nath *et al.* [133], Chen *et al.* [30]) and 3D microscopy images (e.g., Dufour *et al.* [50]). Also for cell tracking active contours are an important and widely used technique (see Sect. 3.4.1 below). However, such methods have the drawback that they are computationally relatively demanding. Related to this type of approaches are the gradient-dependent Voronoi diagrams as described in Jones *et al.* [94] which have been applied for cytoplasm segmentation. There, fronts representing the edges of a Voronoi diagram are evolved towards regions with low local image gradients (i.e. no edges) while the flexibility of the edges is controlled by a regularization parameter.

Overview of our segmentation approach

In this thesis, the task is the segmentation of cell nuclei in 2D fluorescence microscopy images. The images generally exhibit a relatively high signal-to-noise ratio and usually the nuclei are well separated. We have developed a region adaptive thresholding approach that automatically computes local intensity thresholds for subregions of the image. We do not apply additional processing steps to separate clustered nuclei since in our application clustered nuclei represent a phenotype (multinucleated cells) which needs to be quantified. The following sections describe the applied algorithms for automatic threshold computation (Sect. 3.3.2), the region adaptive thresholding approach (Sect. 3.3.3), and finally, extensions of the approach to enable segmentation of morphologically abnormal cell nuclei (Sect. 3.3.5).

3.3.2   Automatic threshold selection

The goal of automatic threshold selection is to find an optimal threshold value that separates the background pixels from the object pixels. Many different thresholding approaches for image segmentation have been described and an extensive overview is given, e.g., in Sezgin *et al.* [162]. There, thresholding techniques have been classified into six categories: (1) histogram shape-based methods, (2) histogram clustering-based methods, (3) entropy-based methods, (4) object attribute-based methods, (5) spatial methods (e.g., correlation), and (6) local adaptive methods. In our approach

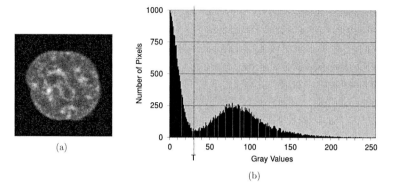

(a)

(b)

Figure 3.4: Example for a gray level histogram. (a) Original image of a cell nucleus, (b) corresponding gray level histogram with threshold T separating the background peak (left) and the foreground peak (right). (For gray value zero the actual histogram value of 12448 has been cut off at 1000 to clearly visualize the two peaks.)

we use two methods from the second category which are *Otsu*'s method [140] and *minimum error thresholding* by Kittler and Illingworth [101]. Both methods perform clustering analysis on the gray level histogram of an image. A gray level histogram represents the frequency of each possible gray value of an image as the number of occurring pixels with the respective gray value. For the case of binary segmentation the goal is to identify two clusters: the histogram peak representing the background and the histogram peak representing the foreground, i.e. the objects of interest (see Fig. 3.4).

Otsu's method

One of the most often used techniques for histogram-based, non-parametric threshold selection is the method of Otsu [140]. This method performs clustering based on the assumption that within each cluster the intensity homogeneity is high, i.e. the values are well clustered around their mean, and, at the same time, the clusters of different classes are well separated. The optimal threshold T_{opt} defining the clusters is determined by maximizing the between-class variance which is equivalent to minimizing the weighted sum of the within-class variances:

$$T_{opt} = argmin\{N_1(T) \cdot \sigma_1^2(T) + N_2(T) \cdot \sigma_2^2(T)\} \tag{3.1}$$

where σ_1^2, σ_2^2 are the variances of both clusters, and the weights N_1, N_2 are the pixel numbers in the clusters, with $N_1(T) = \sum_{z=min}^{T} h(z)$ and $N_2(T) = \sum_{z=T+1}^{max} h(z)$, and

$h(z)$ is the histogram with gray value range $[min, max]$. Thus, this measure takes low values if both clusters have possibly small variances. To find the minimum in (3.1) a sequential search can be performed for all possible values of T.

This method performs well as long as the ratio of background pixels and object pixels N_1/N_2 does not exceed a value of around 10. For larger ratios the segmentation error of Otsu's method increases (Albregtsen [10]).

Minimum error thresholding

The method of Kittler and Illingworth [101] considers the image histogram as a mixture distribution of two Gaussian functions representing the background pixels and the object pixels. The two distributions may have different variances. This method aims to minimize the probability of misclassifying a pixel from one distribution as a pixel from the other distribution. Thus, the overlap of the Gaussian distributions has to be minimized. To this end, Kittler and Illingworth proposed to minimize a criterion function that reflects the average pixel classification error rate:

$$
\begin{aligned}
T_{opt} = \quad & argmin\{1 + 2(N_1(T) \cdot \ln \sigma_1(T) + N_2(T) \cdot \ln \sigma_2(T)) \\
& -2(N_1(T) \cdot \ln N_1(T) + N_2(T) \cdot \ln N_2(T))\}
\end{aligned}
\tag{3.2}
$$

where σ_1, σ_2 are the standard deviations of the two clusters, and N_1, N_2 are the pixel numbers in the clusters, with $N_1(T) = \sum_{z=min}^{T} h(z)$ and $N_2(T) = \sum_{z=T+1}^{max} h(z)$, and $h(z)$ is the histogram with gray value range $[min, max]$. Changing the threshold T changes the model parameters for the two distributions. The better the fit between the data and the modeled Gaussian distributions, the smaller is the overlap of the distributions, and thus, the classification error. The minimum in (3.2) can be determined by evaluating the criterion function for all possible thresholds T. Also, a faster iterative computation is possible but one has to be aware that the criterion function has local minima at the border of the histogram. To avoid the selection of a local minimum, the starting value has to be selected carefully. Using the threshold based on the the approach of *Otsu* as starting value assures convergence towards the minimum error threshold [10].

In comparison to *Otsu*'s method the *minimum error thresholding* approach of Kittler and Illingworth produces satisfying results also if the ratio of background and object pixels N_1/N_2 is above 10^2 [101]. However, note that the estimated model parameters of the Gaussians are always biased as the overlapping tails of the distributions are truncated by histogram partitioning. Thus, the amount of overlap between the two distributions has to be possibly small to yield accurate results.

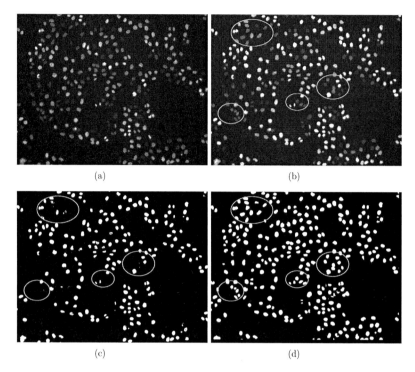

Figure 3.5: Automatic thresholding example. (a) Original image, (b) contrast enhanced original image where regions with low contrast cells are marked by ellipses, (c) segmentation result with global *Otsu* threshold, (d) segmentation result with region adaptive thresholding.

3.3.3   Region adaptive thresholding

Automatic thresholding can produce satisfying results if the gray value ranges of background pixels and object pixels are well separable, i.e. if the gray value histogram has a clear bimodal distribution. However, this might not be the case, e.g., if an image is unevenly illuminated or if an image contains multiple objects of very different intensities. In the latter case, automatic methods such as Otsu's method tend to select the threshold too high, with the result that low contrast cells are classified as background (see Fig. 3.5(a)-(c)). Another issue are closely neighboring objects, where the background between the objects is significantly brighter than in the rest of the image due to scattered light or blurring. Such neighboring objects are often erroneously merged when histogram-based thresholding is applied.

To overcome these problems, region adaptive thresholding can be used. With

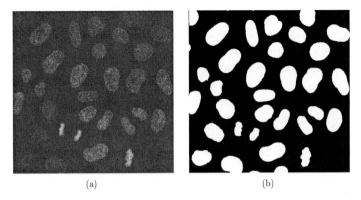

(a) (b)

Figure 3.6: Region adaptive thresholding for maximum intensity projection image from a 3D image sequence. (a) Original image, (b) segmentation result.

region adaptive thresholding an image is split into subregions and local thresholds are computed based on the histograms of the subregions. Thus, the local thresholds are adjusted to local image properties which enables correct segmentation of the problematic cases described above (see Figs. 3.5(d) and 3.6). We have implemented the following scheme for region adaptive thresholding which is based on the approach in Gonzalez *et al.* [64]. For local threshold computation the image is subdivided into quadratic regions which are processed successively. A local threshold is computed only if the variance within a region is above a user-defined threshold, otherwise a global threshold (computed on the whole image) is used. This ensures that a local threshold is calculated only for regions with significant gray value variations and which thus contain a significant amount of information. If the variance is low, i.e. for homogeneous background or object areas the global threshold is applied which reduces the number of local threshold computations and thus, the overall computation time. To calculate global and local intensity thresholds we use the histogram-based threshold selection methods of Otsu [140] and Kittler and Illingworth [101] (see Sect. 3.3.2 above). We found that for different types of images, different combinations of both methods for local and global threshold computation were best suited (see Chapt. 4).

We implemented three alternative strategies to apply the computed local threshold to the image regions: (1) application only to the central pixel of the region, (2) application to the whole region, and (3) application to a quadratic subregion at the center of the region (Fig. 3.7). If the threshold is not applied to the whole region (strategies (1) and (3)), we need to distinguish the outer region in which the threshold is computed, and the inner region in which the threshold is applied. Note that the inner regions need to be adjacent to apply the threshold to the whole image, and consequently, the outer regions overlap. Thus, the total number of image regions as well as the amount of overlap of the regions depends on the used strategy and on the region sizes. Strategy (1) yields the highest accuracies but, on the other hand, requires the most computation time since variance and threshold computations have to be performed for each pixel. As can be seen in Fig. 3.7(a) the regions for threshold computation are highly overlapping. Strategy (2) is fastest but yields segmentation artefacts at the transitions of the non-overlapping adjacent regions (Fig. 3.7(b)). Such artefacts occur if an object lies in two adjacent image regions with highly differing local thresholds (see Fig. 3.8). We evaluated the segmentation accuracy and the computation time for strategies (1) and (2) and compared the results with global thresholding. The evaluation was performed by manually counting correctly and incorrectly segmented cells for four different images (of the same type as shown in Fig. 3.5) that included in total 761 cell nuclei. We found that strategy (1) had the highest accuracy of 98.0% but computation took around 60 sec for one image. Strategy (2), on the other hand, yielded an accuracy of 92.1% but was significantly faster with about 0.2 sec per image. Global thresholding naturally was faster than both strategies with a computation time of 0.09 sec per image, however, this method yielded an accuracy of only 55.9% (see Chapt. 4, Tab. 4.2).

A very good tradeoff between computation time and segmentation accuracy can be achieved with strategy (3) of the algorithm (Fig. 3.7(c)). Here, neighboring regions partly overlap and consequently, adjacent threshold values differ only slightly which minimizes segmentation artefacts. Still, the computation time is much lower than for strategy (1) since complete regions are thresholded at once instead of single pixels. The sizes of the used regions and center regions have to be adapted for different types of images. For the images in our applications we found that a region size of the average nucleus diameter generally produces good results. The center region size has to be chosen according to the desired accuracy and processing speed.

3.3.4   Postprocessing

After thresholding an image, the connected components are determined and labeled. To remove holes within segmented objects a hole filling algorithm is applied. This standard computer graphics algorithm (see, e.g., [83]) first performs flood filling of the background which means to relabel the background as one connected component. Consequently, the remaining regions with the previous background label lie inside of foreground objects and represent the holes to be filled. These regions

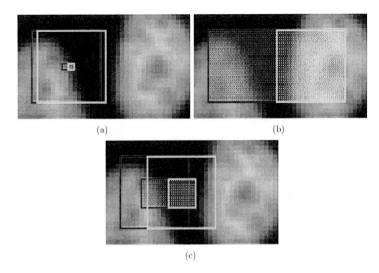

Figure 3.7: Different strategies to subdivide the image into regions for threshold computation and threshold application. In all three cases the threshold is computed in the outer regions. Then the threshold is applied to (a) the center pixel, (b) the whole region, or (c) a center region. The regions to which the thresholds are applied are shaded and in each image two adjacent regions are shown.

are searched and relabeled with the label of the surrounding object. Finally, small objects which are smaller than a user-defined minimum size are removed (scratching). Our segmentation scheme has been implemented in C++ using the generic programming framework of the open-source software ITK [194].

3.3.5 Adaptations for abnormal morphologies

For images of morphologically normal cell nuclei the above described segmentation approach yields satisfying results (see, e.g., Figs. 3.5 and 3.6). However, for images from RNAi knockdown screens, the morphology of the observed cell nuclei can be heavily affected by the treatment. This is particularly true when studying mitosis related genes since a disturbed cell division in many cases also leads to morphologically abnormal daughter cell nuclei. In our image data two types of malformed nuclei appeared which could not be accurately segmented with the above described region adaptive thresholding approach: (1) nuclei with detached chromosomes, and (2) nuclei with attached micronuclei.

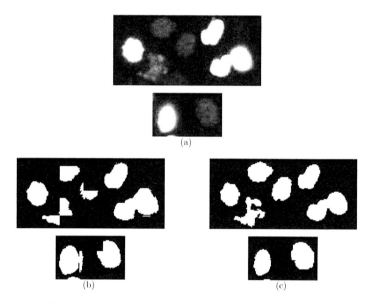

Figure 3.8: Examples for segmentation artefacts. (a) Two sections of an original image (contrast enhanced), (b) corresponding sections of a segmented image using segmentation strategy (2), i.e. application of the threshold to the whole region without region overlap. Typical artefacts are missing rectangular parts for low contrast nuclei and contour artefacts for high contrast nuclei. (c) Corresponding sections of a segmented image using segmentation strategy (1), i.e. application of the threshold to the central pixel with maximally overlapping regions. No artefacts occurred.

Detached chromosomes

Detached chromosomes occur during prometaphase if the formation of the microtubles which build the mitotic spindle is disturbed. In consequence of the disturbed spindle microtubules, single chromosomes are not assembled correctly in the metaphase plate, and thus, are located in some distance to the main chromosome set during cell division and thereafter (see Fig. 3.9(a)). Since in our images the fluorescent dye is attached only to the Histone complex which is located inside of the chromosomes, the spindle is not directly visible. Without visible connection between detached chromosome and mitotic nucleus both structures are segmented as single objects (see Fig. 3.9(b)). However, to determine the correct cell cycle phase of a nucleus it is crucial to treat the main nucleus and its detached chromosome as one object. Consequently, we have developed an algorithm that merges detached chromosomes to mitotic nuclei in their close neighborhood. After image segmentation using the above described approach, detached chromosomes are identified based on their size (number of pixels), which generally is significantly smaller than the average size of the smallest occurring valid objects, that are nuclei in anaphase. Since the sizes of anaphase nuclei vary only in a very small range, a reliable size threshold for detached chromosomes can be determined based on the average size of anaphase nuclei (in our case we used 2000 pixels). For each identified detached chromosome the algorithm determines the closest valid nucleus within a certain Euclidean distance (in our case we used 85 pixels). If there is no nucleus within the search radius the respective chromosome stays unmerged. To find the closest nucleus we use the Euclidean distance of gravity centers instead of the distance between the borders, in order to prevent merging detached chromosomes to interphase nuclei. A nucleus in interphase typically has a roundish shape, and consequently, the distance between gravity center and border has approximately the length of the radius. Since interphase nuclei are generally much larger than mitotic nuclei, a detached chromosome which lies equally close to the border of both, an interphase nucleus and a mitotic nucleus, will always be merged to the mitotic nucleus, as there, the distance of gravity centers is smaller. This prevents mismerging of detached chromosomes even if their borders are closer to an interphase nucleus than to a mitotic nucleus. For merging, the closest pixel of the identified nucleus to the gravity center of the detached chromosome is determined and an artificial line of several pixels width is established between the two points (see Fig. 3.9(c)). After merging, the labeling of the merged objects is updated.

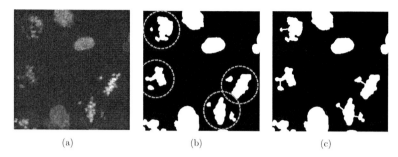

(a)   (b)   (c)

Figure 3.9: Segmentation and merging of detached chromosomes. (a) Original image, (b) segmented image, nuclei with detached chromosomes are encircled, (c) segmented image after applying the merging algorithm.

Attached micronuclei

Nuclei with attached micronuclei occur as a consequence of detached chromosomes during cell division. As the nuclear envelope reforms in telophase, the detached chromosomes form their own nuclear envelopes resulting in micronuclei close to the main nucleus (see Fig. 3.10(a)). Often micronuclei are very dim compared to the neighboring main nucleus. This constellation leads to segmentation artefacts since the automatically determined local threshold is strongly affected by the bright main nucleus. Thus, micronuclei are partly or completely classified as background (see Fig. 3.10(b)). To address this problem, we subdivide the segmentation process into two steps. First, an initial segmentation is determined using the above described approach, which is then used as a mask on the original image. Pixels covered by the mask are set to a constant value (mean intensity multiplied by a constant, in our case we used 1.25) which artificially lowers the intensity of the bright nuclei to the intensity range of the micronuclei (see Fig. 3.10(c)). Second, region adaptive thresholding is performed on the masked image, resulting in a segmentation of the dim objects that were considered as background in the first segmentation run. For the final segmentation result the results of the first and second run are combined by a pixel-wise logical OR operation (see Fig. 3.10(d)). Finally, the postprocessing steps described in Sect. 3.3.4 above as well as the above described procedure for merging of detached chromosomes are performed.

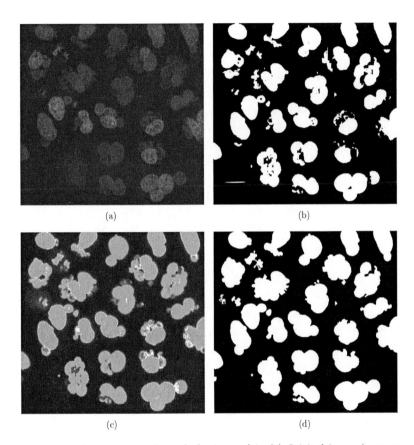

Figure 3.10: Segmentation of attached micronuclei. (a) Original image (contrast enhanced), (b) initial segmentation: result after region adaptive thresholding, (c) original image overlaid with the initial segmentation as a mask (strongly contrast-enhanced), (d) result after region adaptive thresholding of the masked image (c) and combination with the initial segmentation result.

3.4   Tracking of Cell Nuclei

3.4.1   Introduction

To study cell motion, cell development, or temporal changes of cellular features, cells have to be tracked throughout an image sequence. The task of tracking is to establish correspondences between objects in adjacent image frames. This involves strategies to handle ambiguous correspondences resulting from events such as appearing and disappearing objects, as well as splitting and merging objects (one-to-many and many-to-one correspondences). For cell tracking, the handling of splitting objects is particularly important for correctly capturing cell divisions. Most existing approaches for tracking of cells or cell nuclei can be assigned to one of the three main groups of approaches described below.

Deterministic two-step approaches provide a very intuitive strategy to address the tracking problem: first, objects are localized based on segmentation in all images of an image sequence, and second, object correspondences are determined based on object distance (e.g., nearest neighbor) as well as object properties (e.g., geometric or topological properties). Correspondence finding is achieved, for example, based on the Euclidean distance (e.g., Sigal *et al.* [165]), the area overlap of objects (e.g., Gordon *et al.* [65]), or a combination of both (e.g., Withers *et al.* [185], Chen *et al.* [35], Bunyak *et al.* [22]). Additional features such as the volume and the radius of a tracked objects have been used, e.g., in Chen *et al.* [36]. Also, movement properties such as the smoothness of trajectories have been used for correspondence finding (e.g., Chetverikov *et al.* [38]). Initial correspondences can be further refined using graph-based optimization techniques which take into account multiple or all image frames (e.g., Bunyak *et al.* [22], Racine *et al.* [155]).

Deformable models, particularly, active contour-based approaches (e.g., Kass *et al.* [97], McInerney and Terzopoulos [126]), have not only been successfully applied for cell segmentation (see Sect. 3.3.1) but also proved to be well-suited for cell tracking. With this method, the segmentation result obtained in one image frame is used as initialization for the segmentation of the subsequent frame, thereby establishing the correspondence between objects. Generally, one can distinguish between *parametric active contours*, which use explicit parametric representations of object models, and *geometric active contours*, which use implicit representations based on level set functions. Parametric active contours have been applied for 2D cell tracking, for example, in Zimmer *et al.* [202, 203], however, they have the drawback that topological changes (such as splitting or appearing objects) cannot be easily handled. Geometric active contours (e.g., Osher and Sethian [138, 161]), on the other hand, naturally allow contour splits and recognize appearing objects. However, if multiple contours exist in an image, then constraints have to be introduced to prevent contour merging (see below). Geometric active contours have been widely used for cell and cell nucleus tracking in 2D (e.g., Mukherjee *et al.* [131], Zhang *et al.* [198], Yang *et al.* [191], Nath *et al.* [133], Dzyubachyk *et al.* [51]), and have also been extended to active surfaces for tracking cells in 3D image sequences (e.g., Dufour *et al.* [50]). Furthermore, 3D active surfaces have been used for 2D tracking in the spaito-temporal image volume (Padfield *et al.* [141, 143]).

A deformable model based on combined *mean-shift* [37] processes has been described in Debeir *et al.* [47]. There, a set of coupled adaptive kernels has been used to handle changing cell morphologies. Using the mean-shift algorithm with single, non-adaptive kernels typically only allows tracking of simpler morphologies such as blob-like cells (e.g., Bai *et al.* [11]).

Recently, *probabilistic models* have been applied for cell tracking, in particular, stochastic motion filters such as the *Kalman filter* [96] or *particle filters* [66]. A stochastic motion filter represents a tracked object by a state vector, which contains, e.g., its position, speed, or other object features. At each image frame, the motion filter first estimates the state vector for the current frame, based on noisy measurements from previous frames as well as on a motion model. In the second step, the predicted state vector is corrected (updated) based on the new measurements from the current frame which are acquired using a measurement model. In Li *et al.* [113] a Kalman filter, and in Wang *et al.* [182] particle filters have been used for cell tracking in combination with active contour models. Traditional motion filters are limited to only one motion model. To improve tracking of cells with changing motion properties, an *interacting multiple models* (IMM) filter [15] has been used in Li *et al.* [111] including four different motion models.

Mitosis detection and handling is important for live cell tracking. Some approaches naturally enable splitting events (e.g., geometric active contours or graph optimization-based approaches), however, for many tracking approaches additional methods for mitosis detection and track merging have to be included. Level set-based active contour methods naturally allow splits of the evolving front, but, on the other hand, also allow merging of touching cells (which is generally not wanted). To prevent merging, multiple level set functions have been used where each cell is represented by one level set function (e.g., Zhang *et al.* [198], Dufour *et al.* [50]). A computationally more efficient alternative to prevent adjacent cells from merging is to include topological constraints (e.g., Li *et al.* [111, 113]) or to use graph-vertex coloring (e.g., Nath *et al.* [133]). For tracking approaches which do not naturally cope with splitting events, mitosis detection typically is achieved based on geometric and morphological properties of mother and daughter cells, such as the Euclidean distance (Padfield *et al.* [143]), the overlap distance ratio (Withers *et al.* [185]), the difference of sizes (Chen *et al.* [35]) or intensities (Sigal *et al.* [165]). In Padfield *et al.* [141, 144] a sigmoid function was used to model the intensity changes of a nucleus throughout the cell cycle when using a dynamic cell cycle marker, and correspondences between mother and daughter cells were determined based on this model. Also, classification approaches based on supervised learning have been applied for mitosis detection (e.g., Yang *et al.* [191], Li *et al.* [112]).

In this thesis, we present a tracking scheme that can be assigned to the group of *deterministic two-step* approaches. Since processing speed plays an important role for high-throughput applications, computationally more demanding approaches such as *deformable models* are less suited in this context. Our tracking scheme first establishes one-to-one correspondences based on the Euclidean distance as well as the smoothness of the trajectory (Sect. 3.4.2) and thereafter, detects splitting events based on biologically motivated morphological and topological criteria (Sect. 3.4.3).

3.4.2   Correspondence finding

To determine the correspondences of cell nuclei in subsequent image frames we use a feature point tracking algorithm based on Chetverikov *et al.* [38]. As feature points the centers of gravity of segmented cell nuclei are used. For each frame of an image sequence, the algorithm considers the previous and the subsequent image frame. Within these three consecutive frames, object correspondences are established by searching for trajectories with maximum smoothness. For each nucleus the smoothness of trajectories with all potential predecessor and successor nuclei within a limited Euclidean distance is determined and the smoothest trajectory is selected. If $\mathbf{u}$ and $\mathbf{v}$ represent the displacement vectors from the predecessor frame to the current frame and from the current frame to the successor frame, then the following cost function has to be minimized

$$\delta(\mathbf{u}, \mathbf{v}) = w_1 \left(1 - \frac{\mathbf{u} \cdot \mathbf{v}}{||\mathbf{u}|| \cdot ||\mathbf{v}||} \right) + w_2 \left(1 - \frac{2 \cdot (||\mathbf{u}|| \cdot ||\mathbf{v}||)^{1/2}}{||\mathbf{u}|| + ||\mathbf{v}||} \right) \qquad (3.3)$$

where, w_1, w_2 are positive weights. The first term considers the angle θ between adjacent displacement vectors with $\mathbf{u} \cdot \mathbf{v}/||\mathbf{u}|| \cdot ||\mathbf{v}|| = \cos\theta$. Consequently, large angles, i.e. sudden changes of the direction of a trajectory are penalized with higher costs compared to trajectories with smaller angles and thus smoother changes of their direction. The second term takes into account the change of the distance of corresponding feature points by exploiting the ratio of the geometric mean and arithmetic mean of the distances. This ratio decreases if the difference of the distances increases, and it is close to its maximum one if the distances are similar. Hence, trajectories with strongly changing distances cause higher costs than those with relatively smoothly changing distances. The influence of the changes of direction and distance on the result can be controlled by the parameters w_1 and w_2.

3.4.3 Mitosis detection

With the above described approach for correspondence finding only one-to-one correspondences are established. However, if a cell divides, one object in image frame f_{t-1} at time point $t-1$ corresponds to two objects in the subsequent image frame f_t at time point t. In this case, the algorithm links the parent nucleus to the one daughter nucleus with the smoothest trajectory. The other daughter nucleus is treated as an appearing object, i.e. a new trajectory is started. Thus, the information of the relationship of the second daughter nucleus and the parent is not included in the tracking result. To establish correct one-to-many correspondences, a separate method for mitosis detection is performed after correspondence finding. All objects appearing in an image frame f_t (with $t > 0$) after the first image frame f_0 are considered as potential mitosis cases. If a cell division is detected, the respective trajectories are merged. For mitosis detection we first used the overlap distance ratio and later developed a new measure based on morphological criteria of parent and daughter cell nuclei.

The overlap distance ratio

The overlap distance ratio as described in Withers *et al.* [185] takes into account the Euclidean distance of the centers of gravity of the potential parent nucleus and a potential daughter nucleus, as well as the overlap of the areas of both nuclei. Potential daughter nuclei are defined as appearing nuclei within a limited Euclidean distance to the parent nucleus. The scaled distance of the parent cell nucleus i in image frame f_{t-1} and the daughter cell nucleus j in image frame f_t is

$$D_{i,j} = \frac{d_{i,j}}{min_{1 \leq m \leq M, 1 \leq n \leq N}(d_{i,n,t-1}, d_{m,j,t})} \tag{3.4}$$

where $d_{i,j}$ is the Euclidean distance between i and j, M is the total number of cell nuclei in image frame f_{t-1}, and N is the total number of cell nuclei in image frame

f_t. The scaled overlap is defined as

$$V_{i,j} = \frac{B_{i,j}}{min(A_{i,t-1}, A_{j,t})} \qquad (3.5)$$

where A_i and A_j are the areas (number of pixels) of cell nuclei i and j, and $B_{i,j}$ is the amount of overlapping pixels of the two nuclei. In our approach we used the area of the bounding box of a nucleus instead of the nucleus area for reasons explained below (see also [63]). The overlap distance ratio $R_{i,j}$ finally is is given by

$$R_{i,j} = \frac{V_{i,j}}{D_{i,j}} \qquad (3.6)$$

$R_{i,j}$ is always in the interval $[0, 1]$ and close to one if the overlap is large and the distance is small. To decide whether the considered case should be treated as a mitosis event, a threshold for $R_{i,j}$ has to be defined (in our case we used 0.2).

When using the overlap distance ratio it is assumed that after a split the daughter cells do not move far away quickly, and thus, there is a certain overlap of parent and daughter cells. For images of complete cells (i.e. the cytoplasm is also visible) with a sufficiently high image acquisition frequency, this assumption usually is valid. However, in our case we use images of cell nuclei, i.e. only the chromatin and not the cytoplasm is visible. In these image sequences the areas of the parent cell nucleus and daughter cell nuclei do not necessarily overlap. This is because nuclei are relatively small directly before chromosome segregation (usually metaphase) as well as directly afterwards (usually late anaphase) due to high chromosome compaction. Additionally, chromosomes move to the cell poles very quickly and thus, the period in which the areas of parent and daughter nuclei overlap is short. With an image acquisition interval of about 6-7 minutes, this period is not necessarily captured for all cases and consequently, many of the mitosis events are not detected based on the area overlap. Using the bounding box areas instead of the nucleus areas of mother and daughter cell nuclei for the computation of the overlap distance ratio significantly improves the result, but still not all cell divisions are detected (see Fig. 3.11). Based on experimental results we found that this criterion detects 80% of the mitosis cases (see Chapt. 4). Another problem when using the overlap distance ratio is that for images of cell nuclei this measure is prone of generating false positives. In particular, for increasing cell densities, the likelihood of a daughter nucleus having a higher value for the overlap distance ratio with another nucleus than its actual parent rises.

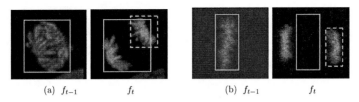

(a) f_{t-1} f_t (b) f_{t-1} f_t

Figure 3.11: Overlap of bounding boxes in two consecutive image frames. Image frame f_{t-1} is before chromosome segregation (solid line bounding box) and image frame f_t afterwards (dashed line bounding box). (a) Bounding boxes do overlap, (b) bounding boxes do not overlap.

Mitosis detection based on morphological criteria

To increase the mitosis detection accuracy, we developed a new approach for mitosis detection that takes into account the morphology as well as the topology of mitotic cell nuclei. As described above, only nuclei that are located within a limited Euclidean distance to the potential parent are considered. The decision for a mitosis event depends on the following two conditions: (1) potential daughter nuclei have to be smaller than the average nucleus size multiplied by a factor c_1 and (2) the Euclidean distance between the potential daughter cells has to be smaller than the average nucleus radius multiplied by a factor c_2 (in our case we used $c_1 = 0.6$ and $c_2 = 3.2$). The average nucleus size and radius are computed based on all nuclei in the respective image sequence. Only if conditions (1) and (2) are fulfilled the potential mitosis case is considered further and a measure for the likelihood L_{i,j_1,j_2} of a mitosis event is computed using

$$
\begin{aligned}
L_{i,j_1,j_2} &= w_1 \left(\frac{1}{2} \cdot \left(1 - \frac{\bar{I}_{all,t}}{I_{j_1,t}}\right) + \frac{1}{2} \cdot \left(1 - \frac{\bar{I}_{all,t}}{I_{j_2,t}}\right) \right) \\
&\quad + w_2 \left(1 - \frac{A_{i,t-1}}{\bar{A}_{all}}\right) \\
&\quad + w_3 \left(\frac{1}{2} \cdot \left(1 - \frac{A_{j_1,t}}{\bar{A}_{all}}\right) + \frac{1}{2} \cdot \left(1 - \frac{A_{j_2,t}}{\bar{A}_{all}}\right) \right) \\
&= w_1 \left(1 - \frac{\bar{I}_{all,t}(I_{j_1,t} + I_{j_2,t})}{2 I_{j_1,t} I_{j_2,t}}\right) \\
&\quad + w_2 \left(1 - \frac{A_{i,t-1}}{\bar{A}_{all}}\right) + w_3 \left(1 - \frac{A_{j_1,t} + A_{j_2,t}}{2\bar{A}_{all}}\right)
\end{aligned}
\tag{3.7}
$$

where j_1, j_2 represent the daughter nuclei in image frame f_t and i the parent nucleus in image frame f_{t-1}. The first term includes the ratio of the mean intensity of all nuclei in the current image frame $\bar{I}_{all,t}$ and the mean intensities of the daughter nuclei $I_{j_1,t}$ and $I_{j_2,t}$. If both daughter cell nuclei are much brighter than the average intensity (which is the case for mitosis), the value of this term is close to one. The second term considers the ratio of the area (i.e. number of pixels) of the parent nucleus $A_{i,t-1}$ and the mean area of all nuclei in the whole sequence $\bar{A}_{all}$. If the parent nucleus is comparatively small, the value of the second term is close to one. The third term finally takes into account the ratio of the areas of the daughter nuclei $A_{j_1,t}$ and $A_{j_2,t}$ and the average nucleus area $\bar{A}_{all}$. Consequently, the value of the third term is close to one if both daughter cell nuclei are small compared to the mean nucleus area. The mean nucleus area is computed based on the whole image sequence to include a possibly large number of cells, in particular, interphase cells. The mean intensity, on the other hand, is computed based only on the current image frame to adapt to temporal illumination fluctuations.

Thus, each term in (3.7) yields values below one, and the values are close to one if mitosis is likely. The terms are weighted by the positive weights w_1, w_2, w_3 (with $w_1 + w_2 + w_3 = 1$). In our application, we used values of $w_1 = 0.5$, $w_2 = 0.2$, $w_3 = 0.3$ which were empirically determined. If L_{i,j_1,j_2} exceeds a threshold (in our case we used 0.2) the case is treated as a mitosis event and the respective trajectories are merged. Using this new mitosis detection criterion, more than 95% of the occurring mitoses were detected correctly compared to manual evaluation with only few false positives (see Chapt. 4).

3.5   Image Features

3.5.1   Introduction

Automatic image classification requires a suitable numeric representation of the image content to be used as input for a learning algorithm. One possibility is to use the gray values of the pixels directly as feature values, possibly after image filtering and segmentation steps. For classification of cell microscopy images such an approach has been used, e.g., in Danckaert *et al.* [43] to classify subcellular structures using neural networks. However, using the gray values as features leads to high-dimensional feature spaces, depending on the resolution of the input image. This increases the computational costs, and reduces the generalization performance for many machine learning methods (curse of dimensionality). In addition, this feature set is not invariant to geometric transformations (translation, rotation, and scaling) of the image content, and thus, provides a very image specific description that only allows recognition of relatively similar images.

Therefore, more general object features are used in most applications for cell image classification. These features describe, e.g., direct object properties such as size, shape or mean gray value, or regional features which take into account the gray value distribution such as texture or moment-based features. An overview of commonly used feature types is given, e.g., in Theodoridis *et al.* [169]. Features for cell image classification have been studied for many years, in particular, for classification of subcellular structures (e.g., Murphy *et al.* [16, 17, 33, 90, 132, 174], Conrad *et al.* [41]). In [17] a standard feature set for subcellular location characterization was introduced and steadily extended to a large set of 2D and 3D image features [90]. Besides classification of subcellular structures also in other fields of microscopy image classification a wide range of image features have been studied (e.g., Ronneberger *et al.* [159], Rodenacker *et al.* [158], Lindblad *et al.* [117], Pincus *et al.* [150]).

Feature extraction based on 3D images: selection of most informative slices

The extraction of image features from 3D images can be either performed by extracting 3D features from the original image stack, or by extracting 2D image features based on single slices or based on a projection of the slices. Here, we use 2D image features since they provide enough information to determine the cell cycle phases (see also Sect. 3.2.2 above). An efficient way to reduce the 3D image feature extraction to a 2D problem is to use the maximum intensity projection of the 3D image stack. The brightest structures of interest of different slices are included in the maximum intensity projection, and, given a low number of slices, the object shape and coarse intensity distribution are still well represented. However, maximum intensity projection leads to blurring of fine textures since the texture can be different or slightly shifted in different image slices. Since texture information is required to distinguish certain cell cycle phases, e.g., interphase and prophase, a loss of texture information should be avoided. Thus, for the extraction of texture features it is advantageous to use original single slices, in particular, the slice containing the maximum information (see Fig. 3.12). To determine this *most informative slice* for each cell nucleus we investigated two measures: the maximum entropy and the maximum total intensity.

The entropy (information entropy or Shannon entropy [163]) $H(g)$ is a measure

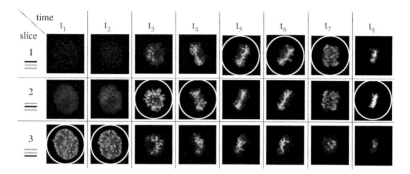

Figure 3.12: Mitotic cell nucleus for eight consecutive time steps (columns) represented by three slices (rows). The white circles denote the *most informative slice* at each time step determined using the maximum total intensity measure. It can be observed that the *most informative slice* changes from the bottom slice to the top slice during mitosis.

for the information of an image g and is defined as

$$H(g) = -\sum_{i=0}^{N_g-1} p(g_i) \cdot \log_2 p(g_i) \qquad (3.8)$$

where g_i are the gray values occurring in image g, N_g is the number of gray values, and $p(g_i)$ is the relative frequency of gray value g_i, $i \in [0, N_g\text{-}1]$ (the log is taken to base 2, as it is usually done to compute the entropy in bits). To find the most informative slice we determined the slice with the maximum entropy. However, we found that this measure is relatively sensitive to noise for our image data. Selecting the slice with the maximum total intensity, on the other hand, yielded very good results. The total intensity $I(g)$ of an image g can be written as

$$I(g) = \sum_{i=0}^{N_g-1} g_i \cdot p(g_i) \qquad (3.9)$$

An example for most informative slice selection for the two measures is given in Fig. 3.13. In this example, the maximum entropy results for the top slice (number three) which does not represent the object well. The maximum total intensity is yielded for the bottom slice (number one) which actually provides the best representation of the object.

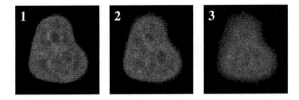

Figure 3.13: Three slices of a typical interphase cell nucleus (1=bottom, 2=middle, 3=top). The maximum entropy results for the top slice (no. 3), while the maximum total intensity is yielded in the bottom slice (no. 1).

3.5.2 Static features

For the computation of static image features we adopted a large feature set which has been previously used for the classification of subcellular phenotypes in [40]. This feature set includes object- and edge-related features, granularity features, tree-structured wavelet features, Haralick texture features, gray scale invariants, and moment based features. We extended this feature set by shape features (circularity, contour length, Feret's diameter). In total, we compute more than 350 features per cell object.

Object- and edge-related features

This group of features describe basic properties of an object, like the area (number of pixels), the contour length (i.e. the perimeter), the mean gray value, and the standard deviation of gray values. Also, basic shape features such as the circularity and Feret's diameter are used. The circularity c of an object is computed by

$$c = \frac{p^2}{A} \qquad (3.10)$$

were p is the perimeter of the object and A is its area. The minimum value for c is 4π in the case of a circle. Thus, c can be normalized with 4π so that a circle has a circularity of one. Maximum Feret's diameter (caliper length) is defined as the greatest possible distance between any two contour pixels. Furthermore, we use edge-related features which are computed by applying Laplace and Sobel filters to the image and refining the detected edges with thresholding. As feature values we use the number of detected edge pixels.

Haralick texture features

The Haralick texture features [72] are computed based on so called co-occurrence matrices (or spatial dependence matrices) of an image. A co-occurrence matrix represents the relative frequencies of all occurring gray value pairs of pixels at a given distance d with the angle ϕ. For an image with gray values in the range of $[0, N_g\text{-}1]$ the co-occurrence matrix Co for a pair (d, ϕ) is defined as

$$
Co(d, \phi) = \frac{1}{R}
\begin{bmatrix}
\eta(0,0) & \eta(0,1) & \cdots & \eta(0, N_g\text{-}1) \\
\eta(1,0) & \eta(1,1) & \cdots & \eta(1, N_g\text{-}1) \\
\vdots & \vdots & \ddots & \vdots \\
\eta(N_g\text{-}1, 0) & \eta(N_g\text{-}1, 1) & \cdots & \eta(N_g\text{-}1, N_g\text{-}1)
\end{bmatrix}
$$

were $\eta(g_i, g_j)$ is the frequency of gray value pair (g_i, g_j) and R is the total number of possible pixel pairs in the image (depending on d and ϕ). Hence, $P(g_i, g_j) = \frac{1}{R}\eta(g_i, g_j)$ is the probability for a gray value pair (g_i, g_j). Here, we compute co-occurrence matrices for the distances of one to five pixels and angles of 0°, 45°, 90°, and 135°. For each co-occurrence matrix 13 features (second order-statistics, e.g., angular second moment, contrast, correlation, variance, entropy, etc.) are computed leading to 260 image features that describe the texture of an image (see Tab. 3.1). As an example consider the angular second moment f_1 which is a measure of smoothness. If all image pixels have same gray value k (i.e. maximum smoothness) then $P(k, k) = 1$ and $P(i, j) = 0 \; \forall \; i, j \neq k$, and the angular second moment yields its maximum value with $f_1 = 1$. If all pixel pairs have equal probability $\frac{1}{R}$ (random distribution), then $f_1 = \frac{R}{R^2} = \frac{1}{R}$. Thus, the less smooth an image is, the lower is its value for f_1.

Granularity features

We compute granularity features which also describe texture based on the relation of neighboring pixel pairs. Here, the gray value differences of a center pixel to all pixels within a given distance in eight directions (0°, 45°, 90°, 135°, 180°, 225°, 270°, and 315°) are computed and the maximum difference is determined. As features we use the mean and the standard deviation of the maxima calculated over all image pixels [40]. The granularity features are computed for distances of one to ten pixels.

Tree-structured wavelet features

The wavelet transform decomposes a signal into different frequency channels. The signal is decomposed using a family of wavelets obtained by scaling a function $\psi(t)$

Angular Second Moment

$f_1 = \sum_i \sum_j (P(i,j))^2$

Sum Entropy

$f_8 = -\sum_{i=0}^{2N_g-2} P_{x+y}(i) \log P_{x+y}(i)$

Contrast

$f_2 = \sum_{n=0}^{N_g-1} n^2 \left\{ \begin{array}{c} \sum_i \sum_j P(i,j) \\ |i-j|=n \end{array} \right\}$

Entropy

$f_9 = -\sum_i \sum_j P(i,j) \log P(i,j) \equiv H_{xy}$

Correlation

$f_3 = \frac{\{\sum_i \sum_j (ij) P(i,j)\} - \mu_x \mu_y}{\sigma_x \sigma_y}$

Difference Variance

$f_{10} = \sum_{i=0}^{N_g-1} (i - \hat{f}_6)^2 P_{x-y}(i)$

Variance

$f_4 = \sum_i \sum_j (i - \mu)^2 P(i,j)$

Difference Entropy

$f_{11} = -\sum_{i=0}^{N_g-1} P_{x-y}(i) \log P_{x-y}(i)$

Inverse Difference Moment

$f_5 = \sum_i \sum_j \frac{P(i,j)}{1+(i-j)^2}$

Information Measure I

$f_{12} = \frac{H_{xy} - H_{xy}^1}{\max\{H_x, H_y\}}$

Sum (Difference) Average

$f_6(\hat{f}_6) = \sum_{i=0}^{2N_g-2(N_g-1)} i P_{x+(-)y}(i)$

Information Measure II

$f_{13} = \sqrt{1 - \exp(-2(H_{xy}^2 - H_{xy}))}$

Sum Variance

$f_7 = \sum_{i=0}^{2N_g-2} (i - f_6)^2 P_{x+y}(i)$

Definitions

$H_{xy}^1 = -\sum_i \sum_j P(i,j) \log(P_x(i) P_y(j))$

$P_x(i) = \sum_j P(i,j)$

$P_{x\pm y}(k) = \sum_i \sum_{j, |i\pm j|=k} P(i,j)$

$H_{xy}^2 = -\sum_j \sum_i P_x(i) P_y(j) \log(P_x(i) P_y(j))$

$P_y(j) = \sum_i P(i,j)$

$\mu, \mu_x, \mu_y, \sigma_x, \sigma_y; H_x, H_y$

means, std. deviations and entropies.

Table 3.1: Haralick texture features as given in Theodoridis *et al.* [169].

(called the *mother wavelet*) by s and translating it by u [120]:

$$\psi_{u,s}(t) = \frac{1}{\sqrt{s}}\psi\left(\frac{t-u}{s}\right) \qquad (3.11)$$

The *continuous* wavelet transform of a function $f(t)$ is defined as

$$w_{u,s} = \int_{-\infty}^{+\infty} f(t)\psi_{u,s}(t)dt \qquad (3.12)$$

and provides the wavelet coefficients $w_{u,s}$ which allow reconstructing $f(t)$ using the synthesis formula

$$f(t) = \sum_{u,s} w_{u,s}\psi_{u,s}(t). \qquad (3.13)$$

The *discrete* wavelet transform (DWT) in practice can be computed using discrete convolutions of the input signal with a low-pass filter l (scaling filter) and a high-pass filter h (wavelet filter) which together define the wavelet. This twofold filtering results in a decomposition of the input signal into a set of *approximating coefficients* a_k as the output of the low-pass filter and a set of *detail coefficients* d_k as the output of the high-pass filter. In a multi-resolution approach this filtering is repeated recursively on the approximating coefficients, where each decomposition step is combined with a downsampling of the input coefficients by two. The output of the high-pass filter (detail coefficients) is determined for each resolution level j but not further processed. The approximating coefficients $a_{j+1,k}$ and the detail coefficients $d_{j+1,k}$ at resolution level $j+1$ can be computed based on the approximating coefficients $a_{j,n}$ from the previous resolution level j using

$$a_{j+1,k} = \sum_n a_{j,n}\, l(n-2k) = a_j * l(-2k) \qquad (3.14)$$

$$d_{j+1,k} = \sum_n a_{j,n}\, h(n-2k) = a_j * h(-2k). \qquad (3.15)$$

This successive subdivision of the low frequency channel is also called *pyramid-structured* wavelet transform. The DWT of a 2D input signal (e.g., an image) can be obtained by sequentially transforming the rows and the columns of the input matrix. Given an image this results in four subimages, where each subimage comprises a part of the whole frequency bandwidth and has a quarter of the input image resolution (see Fig. 3.14(a)). Here, we use Daubechies wavelets [46] which are widely used in signal and image processing. In particular, we use a 12-tap Daubechies wavelet, i.e. h and l are defined based on 12 coefficients.

The idea of the *tree-structured* wavelet transform [28] is to decompose further

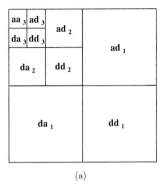 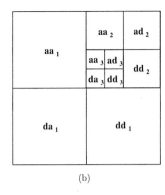

(a)   (b)

Figure 3.14: Decomposition schemes for the discrete wavelet transform of a 2D signal (e.g., an image) for three resolution levels ($j = 3$). (a) Pyramid-structured wavelet transform: only the low frequency channel that contains the approximating coefficients for the row and the column transformation aa_j is decomposed further. (b) Tree-structured wavelet transform: the frequency channel with the highest energy is subdivided. In this example the highest energy channels are ad_1, da_2.

the frequency channels that contain the most information instead of decomposing the lowest frequency channel as for the *pyramid-structured* wavelet transform (see Fig. 3.14). The information content is determined using the image energy. Here, the mean of the absolute gray values is used as energy function $E(g)$ for an image $g(x,y)$ with x- and y-dimensions N_x, N_y:

$$E(g) = \frac{1}{N_x N_y} \sum_{x=0}^{N_x-1} \sum_{y=0}^{N_y-1} |g(x,y)| \tag{3.16}$$

The input image is decomposed recursively several times depending on the image size. As features we compute at each decomposition step the product of the highest energy value and a constant representing the frequency channel in which the highest energy occurred.

Gray scale invariants

Gray scale invariants represent structural information independently of the object's position and orientation [24]. Gray scale invariant features are computed by combining a pair of neighboring pixels in an image $g(x,y)$ using a kernel function f of the type

$$f(g(x,y)) = f_1(g(x,y)) \cdot f_2(g(x+d_1, y+d_2)) \tag{3.17}$$

where f_1 and f_2 are arbitrary functions that transform the gray values, and $\mathbf{d} = [d_1\ d_2]^T$ is the span vector of the kernel function. This function is computed for a pixel and its neighbors in all possible directions and the results are summed up, providing a value that is invariant to rotation of the image content. The set of rotated kernel functions is evaluated at all possible positions of the image and the results are summed up over the whole image, providing a value that is invariant to translation of the image content. This can also be formulated as a convolution (denoted as $*$) of the image with a ring mask M, followed by integration:

$$T_f(g) = \frac{1}{2\pi N_x N_y} \sum_{x=0}^{N_x-1} \sum_{y=0}^{N_y-1} F_1(x,y) \cdot (F_2 * M)(x,y) \tag{3.18}$$

where $F_1 = f_1(g)$ and $F_2 = f_2(g)$, N_x, N_y are the image dimensions in x- and y-direction, and $M(x,y)$ is a ring mask with values of zero in the background and values above zero defining the ring with a radius of $||\mathbf{d}||$. The applied kernel functions are (1) the product of the gray values (i.e. $f_1(g) = f_2(g) = g$) and (2) the product of the square roots of the gray values (i.e. $f_1(g) = f_2(g) = \sqrt{g}$). Different gray scale invariant features can be computed using different radii $||\mathbf{d}||$ for the kernel function.

Moment-based features

Moments are used in statistics to characterize distributions of random variables, as well as in mechanics to characterize bodies by their spatial distribution of mass. Similarly, moments can be used in image processing by considering an image region as a 2D density distribution function. Moments of different orders and with different basis functions are used in image analysis to describe information contained in image regions (Prokop et al. [152]).

Geometric moments   The two-dimensional Cartesian moment m_{pq} of order $p+q$ of an image $g(x,y)$ is

$$m_{pq} = \sum_{x=0}^{N_x-1} \sum_{y=0}^{N_y-1} x^p y^q g(x,y). \tag{3.19}$$

Here, the basis function is the monomial product $x^p y^q$. The translation invariant central moments μ_{pq} are computed using

$$\mu_{pq} = \sum_{x=0}^{N_x-1} \sum_{y=0}^{N_y-1} (x - \bar{x})^p (y - \bar{y})^q g(x,y) \tag{3.20}$$

where

$$\bar{x} = \frac{m_{10}}{m_{00}}, \bar{y} = \frac{m_{01}}{m_{00}}.$$

$\bar{x}$ and $\bar{y}$ represent the coordinates of the center of gravity. We determine the second order moments (moments of inertia) which provide information, e.g., on the principal axes of an object. As feature value we use the ratio of the second order central moments μ_{20} and μ_{02}.

Zernike moments Zernike moments use complex Zernike polynomials [195] as moment basis functions and enable the computation of image features that are invariant to translation and rotation. Since Zernike polynomials are defined within the unit circle, the image has to be translated and scaled to a unit disc first (disc of radius one, centered at the origin (0,0)). The Zernike moments Z_{nl} then can be computed for an image $g(x, y)$ using

$$Z_{nl} = \frac{n+1}{\pi} \sum_{x=0}^{N_x-1} \sum_{y=0}^{N_y-1} V_{nl}^*(x, y) g(x, y) \qquad (3.21)$$

with $x^2 + y^2 \leq 1$ and $V_{nl}^*(x, y)$ is the complex conjugate of a Zernike polynomial of degree n and

$$V_{nl}(x, y) = \sum_{m=0}^{(n-l)/2} (-1)^m \frac{(n-m)!}{m! \left[\frac{(n-2m+l)}{2}\right]! \left[\frac{(n-2m-l)}{2}\right]!} \cdot (x^2 + y^2)^{\frac{n}{2}-m} e^{il\theta} \qquad (3.22)$$

where l is a positive integer with $0 \leq l \leq n$, n-l is even, and $\theta = \tan^{-1}(y/x)$, $i = \sqrt{-1}$. We use the moment's magnitudes $|Z_{nl}|$ as image features as proposed, e.g., in Boland *et al.* [16]. The magnitudes of the Zernike moments are invariant to rotation as it has been shown in Khotanzad *et al.* [98]. We calculate the Zernike moments up to degree 12 using all possible values for l which results in 49 features.

3.5.3 Dynamic features

In addition to the static image features described above we developed dynamic features which represent changes of the shape and intensity of cell nuclei in consecutive image frames. To this end, we use the ancestral information that has been determined by the tracking algorithm. To determine dynamic features we compute the difference of basic image features for each cell nucleus at time point t to the corresponding nuclei at the previous time point $t - 1$, and the subsequent time point

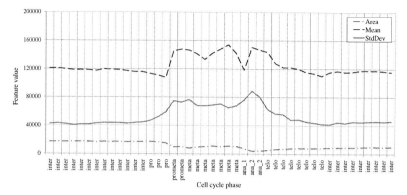

Figure 3.15: Plot of three basic features over time determined for one representative cell nucleus. The displayed features are (from top to bottom) the mean intensity, the standard deviation of intensities, and the area (number of pixels). The curves exhibit characteristic profiles which can be used to describe the morphological changes of a cell nucleus throughout the cell cycle.

$t + 1$:

$$f_{dyn}(t) = f_{stat}(t) - f_{stat}(t \pm 1) \tag{3.23}$$

The basic features we use are object size, mean intensity, standard deviation of intensities, circularity and Feret's diameter. For these features we observed prominent and characteristic changes during mitosis. As an example, Fig. 3.15 shows a plot of the area, the mean intensity and the standard deviation of intensities over time for one mitotic cell. For the mean intensity it can be seen that, e.g., between *prophase* and *prometaphase* there is a steep rise, and for *early anaphase* there is a characteristic spike. The standard deviation of intensities exhibits a similar profile. The temporal change of the area is also characteristic: the area generally increases until the beginning of *prometaphase*, then decreases until *late anaphase* and afterwards steadily increases again. To determine dynamic features also other types of features could be used. We measured the benefit of these dynamic features by means of classification accuracy. In Sect. 4.3 we demonstrate that the inclusion of dynamic features can improve the classification results significantly compared to using only static features. In particular, for object classes with few training samples and high intra-class variability we show that the dynamic features increased the classification accuracy significantly.

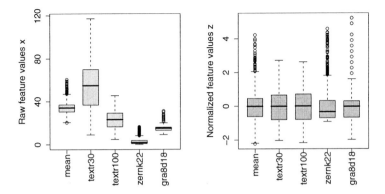

Figure 3.16: Feature value distribution for five exemplary features based on 1264 cell nuclei from one image sequence. The features are (from left to right): the mean intensity, two different Haralick texture features, a Zernike moment, a granularity feature. The diagram on the left shows the distribution before feature normalization, the diagram on the right provides the distribution after normalization. The line within a box represents the median, the upper and the lower border of the box represent the quantiles, the circles are outliers (distance to median larger than 1.5 interquantile ranges).

3.5.4 Feature postprocessing

Feature normalization

Before the features are used for classification, the feature values usually are normalized over the whole population of samples. By normalization, the values of different features are transformed into approximately the same numeric range which avoids that features having a larger numeric range dominate those with a smaller range and thus improves the classification performance (Hsu *et al.* [87]). For normalization, we apply the standard z-score method. All samples x_i within one class of features are transformed to normalized values z_i using

$$z = \frac{x - \mu}{\sigma} \tag{3.24}$$

where μ is the mean value of the feature values x_i and σ is the standard deviation. Consequently, the transformed feature values z_i have a mean value of zero and unit variance. An example illustrating the effect of feature normalization is shown in Fig. 3.16 for five arbitrary features. The z-score can also be used to normalize features between different experiments.

Feature reduction

Feature reduction is required for many classifiers to prevent overfitting to the training data. This is particularly important if the number of training samples is low
compared to the dimensionality of the feature space (*curse of dimensionality*). However, kernel-based methods such as support vector machines do not suffer from this
problem (Burges [23]) and thus, the classification accuracy generally is not affected
in case of high dimensional feature spaces. Still, feature reduction is preferable since
the computation time for the training of the classifier is lower for feature spaces
of lower dimensionality. Also, a lower number of features reduces the computation
time for feature extraction. However, here we have to distinguish methods for feature space reduction which generate new, uncorrelated features as a recombination
of the original features, and methods for feature selection which select a feature
subset out of the original feature set. Only in the latter case the total number of
computed features can be generally reduced. In the first case, all original features
have to be computed to generate the reduced set of uncorrelated features. Different
feature reduction strategies have been compared for biological images, e.g., in Huang
et al. [91], Wang *et al.* [180], and Kovalev *et al.* [103].

Feature recombination   In the context of this thesis we applied two methods
typically used for feature reduction, namely principal component analysis (PCA, discrete Karhunen-Loève transform) [93], and independent component analysis (ICA)
[39, 92].

PCA transforms the original feature values into a new coordinate system of lower
dimensionality which is spanned by a subset of m principal components of the data.
The principal components are defined by the eigenvectors of the data's covariance
matrix. As basis vectors for the new coordinate system the eigenvectors with the
highest eigenvalues are selected. The m largest eigenvalues correspond to the components that contribute most to the total variance of the dataset. Nevertheless,
neglecting the eigenvectors with low eigenvalues means a loss of information. Thus,
to determine a suitable value for m a tradeoff has to be found between the degree of
dimensionality reduction and the loss of information. In essence, PCA retains the
components with the largest variances which are supposed to be the most relevant
for class separation. Note that a Gaussian distribution of the original features is
assumed.

ICA decomposes a given multivariate compound signal into its statistically independent additive source signals. This decomposition is based on the assumption
that the independent source components have a less Gaussian character than the

additive compound signal. To maximize the statistical independence of m source components, the mutual information between the source components is minimized. Here, the mutual information is defined based on the negentropy. The negentropy $J(\mathbf{y})$ for a random vector $\mathbf{y} = (y_1 \ldots y_m)^T$ (representing the source components) is the difference of the entropy $H(\mathbf{y}_{Gauss})$ and $H(\mathbf{y})$, where $\mathbf{y}_{Gauss}$ is a Gaussian random vector with the same covariance matrix as $\mathbf{y}$ (for a definition of the entropy H see (3.8) above)[39, 92]. Thus, the negentropy can be interpreted as a measure of non-Gaussianity, and the mutual information MI is given as [39, 92]

$$MI(y_1, y_2, \ldots, y_m) = J(\mathbf{y}) - \sum_{i=1}^{m} J(y_i). \qquad (3.25)$$

In (3.25) small values for MI are yielded if the sum of the negentropies of the components y_i is large compared to the negentropy of the compound signal $\mathbf{y}$. In other words, ICA aims to find *interesting* projections of the joint distribution of the data which provide estimates of the independent components.

ICA is related to PCA, though under the contradictory assumptions of non-Gaussianity and Gaussianity, respectively. ICA is the more powerful technique, capable of finding the underlying sources even if PCA fails. However, most ICA methods are not able to determine the actual number of components. Thus, often PCA is used in the first step for selecting a suitable number of components to be later extracted by ICA. In our case, the application of PCA resulted in comparable classification accuracies as using ICA with the same number of components (see Sect. 4.2 and 4.4).

Feature selection Various methods for feature selection have been described in the literature and an overview is given, e.g., in Dash *et al.* [44]. Feature selection strategies usually involve a method for feature set generation (search procedure) as well as an evaluation function (selection criterion) [14]. The optimal generation procedure is an exhaustive search, i.e. $N = \binom{n}{x}$ combinations have to be tested to select a subset of x features from a feature set of size n. For larger feature sets this strategy is computationally not feasible and thus, usually suboptimal search strategies are applied (see, e.g., Dash *et al.* [44], Theodoridis *et al.* [169], Bishop [14]). The evaluation function is required to measure the quality of a tested feature or feature subset. As evaluation functions either statistical methods, e.g., hypothesis testing or class separability measures (e.g., distance measures) are applied to the features, or the increase of the classification accuracy for the tested feature subset is used as criterion. We tested two methods based on feature statistics, namely the stepwise discriminant analysis which is based on a Fisher test with Wilks' λ (Leray *et al.* [108]), and SAM (significant analysis of microarrays) which involves a permutation test on a modified T-test statistic (Tusher *et al.* [171]). Both methods were applied using the data mining platform Mine-It [4]. Furthermore, we tested a method based on the classification accuracy (Puig *et al.* [153]). This straightforward approach first generates a ranked feature list based on the single feature classification accuracies and then sequentially adds features from the top of the list to the optimal subset. At each iteration the classification accuracy for the current feature subset is determined using cross validation (see Sect. 3.6.3). The process can be stopped as soon as the desired classification accuracy is reached. Experimental results of the performance evaluation of these feature selection methods are provided in Sect. 4.3 below.

3.6   Classification of Cells

3.6.1   Introduction

Classification methods generally can be distinguished into supervised and unsupervised learning methods. Supervised learning requires training of the classifier with a set of correctly classified examples and allows classification into the predefined classes only. Unsupervised learning (also denoted as clustering), on the other hand, clusters the data depending on the data distribution in the feature space without predefined classes. In applications where the expected classes are previously known, higher classification accuracies can usually be achieved using supervised learning methods. Previous approaches on classification of cell microscopy images, in most cases applied supervised learning methods since there, the classes are usually defined by the application. The most commonly used supervised classification methods for cell microscopy images are neural networks (e.g., Murphy *et al.* [16, 17, 33, 132, 174], Danckert *et al.* [43], Conrad *et al.* [41], Nattkemper *et al.* [135]), support vector machines (SVM) (e.g., Conrad *et al.* [41], Neumann *et al.* [136, 177], Murphy *et al.* [32], Hamilton *et al.* [69]), k-nearest neighbor classifiers (KNN) (e.g., Würflinger *et al.* [189], Wong *et al.* [35, 178]), Bayesian classifiers (e.g., Würflinger *et al.* [189], Wong *et al.* [178]), linear discriminant analysis (LDA) (e.g., Lindblad *et al.* [117], Wong *et al.* [178]), hidden Markov models (e.g., Gallardo *et al.* [57], Wong *et al.* [180, 200]), and ensemble methods (e.g., Murphy *et al.* [89, 90], Jones *et al.* [95]). Comparisons between different classification methods have been performed based on cell microscopy images, e.g., of subcellular structures (Conrad *et al.* [41], Murphy *et al.* [89, 90]), cell nuclei (Kovalev *et al.* [103], Wong *et al.* [180]) or complete cells (Wong *et al.* [178]).

However, for applications where the classes are not previously known, unsupervised learning methods can be used to cluster the data. Clustering methods have been applied, e.g., to determine protein relations based on similar location patterns (Murphy *et al.* [34, 90]). Unsupervised learning methods can also be applied for the analysis of genome-wide screening experiments. In such experiments usually a wide range of phenotypes occur which are not necessarily known beforehand, and consequently, a supervised classification method (trained only on known phenotypes) cannot classify them correctly. Using clustering techniques to discover new phenotypes in large data sets has been proposed, e.g., in [177, 178].

In this thesis, we use a supervised learning method for cell nucleus classification since in our application the analysis task is focused on previously known classes. In particular, we use support vector machines (SVMs) to classify cell nuclei in different cell cycle phases. This classifier showed a very good classification performance for our data, and thus, proved to be well suited for our application. However, we also investigated other supervised methods (hierarchical clustering [21], random forests [20]) and unsupervised methods (k-means clustering, hard competitive learning, neural gas [48]) on our data and compared the results (see Sect. 4.2). We found that the best results were achieved using SVMs or random forests.

3.6.2 Support vector machines

Support vector machines (SVMs, Vapnik [172, 173]) have become a popular and widely used method for classification in the field of pattern recognition. In the following, we describe the basic concepts of SVMs (see Burges [23], Chen *et al.* [31]). A support vector machine computes an optimal separating hyperplane between the class samples of the training data in the feature space. This hyperplane is optimal in the sense that its distance to the nearest training data points, i.e. the margin, is maximized (see Fig. 3.17(a)).

Linear support vector machines

The separable case A binary (two-class) classification problem with linearly separable data can be formulated as follows. As training samples consider pairs $\{\mathbf{x}_i, y_i\}$, $i = 1, \ldots, l$, where $\mathbf{x}_i \in \mathbf{R}^d$ is the feature vector and $y_i \in \{-1, 1\}$ is the respective class. The separating hyperplane is defined by the points $\mathbf{x}$ satisfying $\mathbf{w} \cdot \mathbf{x} + b = 0$, where $\mathbf{w}$ is a normal vector of the hyperplane and $|b|/\|\mathbf{w}\|$ is the perpendicular distance of the hyperplane to the origin. The data points that are closest to the hyperplane define the margin and are called support vectors. The margin thus is represented by two parallel hyperplanes H_1, H_2 with equal distance to the separating hyperplane (Fig. 3.17(a)). Suppose that all training samples satisfy the constraints

$$\mathbf{x}_i \cdot \mathbf{w} + b \geq +1 \quad for \quad y_i = +1 \tag{3.26}$$

$$\mathbf{x}_i \cdot \mathbf{w} + b \leq -1 \quad for \quad y_i = -1. \tag{3.27}$$

Then the margin is given by the distance of the hyperlanes $H_1 : \mathbf{x}_i \cdot \mathbf{w} + b = 1$ and $H_2 : \mathbf{x}_i \cdot \mathbf{w} + b = -1$. The perpendicular distances of H_1 and H_2 to the origin are $|-b-1|/||\mathbf{w}||$ and $|-b+1|/||\mathbf{w}||$, and subtracting both distances results in the value of the margin $2/||\mathbf{w}||$. Thus, to determine the pair of hyperplanes (H_1, H_2) that provide the maximum margin we need to minimize $||\mathbf{w}||$ subject to the inequality constraints

$$y_i(\mathbf{x}_i \cdot \mathbf{w} + b) - 1 \geq 0 \quad \forall i \tag{3.28}$$

which are obtained by combining the constraints (3.26) and (3.27). Since it is difficult to solve this optimization problem for $||\mathbf{w}||$ (as the norm includes a square root), $||\mathbf{w}||$ is substituted by $\frac{1}{2}||\mathbf{w}||^2$ which does not change the results for $\mathbf{w}$ and b, and simplifies the solution. To find extrema of a multivariate function $f(\mathbf{w}, b)$ subject to a set of constraints, the method of Lagrange multipliers can be used. The here described minimization problem is solved by introducing positive Lagrange multipliers $\alpha_i, i = 1, \ldots, l$. With the function $f(\mathbf{w}, b) = \frac{1}{2}||\mathbf{w}||^2$ to be minimized and the constraints $c(\mathbf{w}, b)$ given in (3.28) the Lagrangian is

$$L_P \equiv \frac{1}{2}||\mathbf{w}||^2 - \sum_{i=1}^{l} \alpha_i y_i(\mathbf{x}_i \cdot \mathbf{w} + b) + \sum_{i=1}^{l} \alpha_i, \quad \alpha_i \geq 0 \quad \forall i \tag{3.29}$$

Next, the partial derivatives $\frac{\partial}{\partial w} L_P$ and $\frac{\partial}{\partial b} L_P$ are set to zero which results in the following conditions

$$\mathbf{w} = \sum_{i=1}^{l} \alpha_i y_i \mathbf{x}_i \tag{3.30}$$

$$\sum_{i=1}^{l} \alpha_i y_i = 0. \tag{3.31}$$

In (3.29) L_P represents the so-called primal formulation of the optimization problem, however, a complementary, so-called dual formulation L_D of the problem can be obtained by substituting (3.31) and (3.30) into (3.29). Both formulations are based on the same objective function but with different constraints, and the same solution is obtained by minimizing L_P as by maximizing L_D. The dual formulation of the problem is

$$L_D = \sum_{i=1}^{l} \alpha_i - \frac{1}{2} \sum_{i,j=1}^{l} \alpha_i \alpha_j y_i y_j \mathbf{x}_i \cdot \mathbf{x}_j. \tag{3.32}$$

Note that in the dual formulation the data appears in form of the dot product $\mathbf{x}_i \cdot \mathbf{x}_j$. This is an important property for the realization of non-linear SVMs, where the dot product is replaced by a non-linear kernel function (see (3.39) below). For training of the SVM, L_D is maximized with respect to α_i subject to constraints (3.31) and positivity of the α_i. Note that for each training sample there is a Lagrange multiplier α_i where the cases with $\alpha_i > 0$ are the support vectors. For all other training samples $\alpha_i = 0$. Finally, $\mathbf{w}$ is given by (3.30) and b can be determined using the *complementary* condition of the *Karush-Kuhn-Tucker* (KTT) conditions for the primal problem L_P [23, 31, 56]:

$$\alpha_i (y_i(\mathbf{w} \cdot \mathbf{x}_i + b) - 1) = 0 \quad \forall i \tag{3.33}$$

To apply the trained SVM for classification of a test sample $\mathbf{x}$, the following hyper-plane decision function has to be evaluated

$$f(\mathbf{x}) = sgn(\mathbf{w} \cdot \mathbf{x} + b) = sgn \left(\sum_{i=1}^{l} \alpha_i y_i (\mathbf{x} \cdot \mathbf{s}_i) + b \right) \tag{3.34}$$

where $\mathbf{s}_i$ are the support vectors. Again, the data appears in form of a dot product $\mathbf{x} \cdot \mathbf{s}_i$.

The non-separable case For non-separable data additional positive slack variables ξ_i are introduced which relax the set of constraints (3.28):

$$y_i(\mathbf{x}_i \cdot \mathbf{w} + b) \geq 1 - \xi_i \quad \forall i, \tag{3.35}$$
$$\xi_i \geq 0 \quad \forall i.$$

This formulation allows that training samples may lie on the wrong side of the decision hyperplane (see Fig. 3.17(b)) and realizes a *soft margin* classifier. The sum of the slack variables $\sum_i \xi_i$ provides an upper bound on the number of training errors, and the objective function to be minimized changes from $\frac{1}{2}||\mathbf{w}||^2$ to $\frac{1}{2}||\mathbf{w}||^2 + C \sum_i \xi_i$, where the parameter C regulates the penalty of errors and has to be chosen by the user. Introducing the Lagrangian multipliers α_i and μ_i to formulate the primal

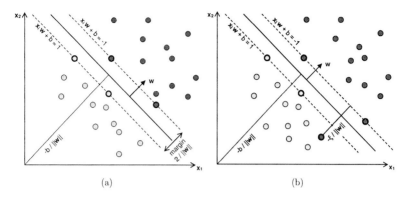

                              (a)   (b)

Figure 3.17: Linear separating hyperplane with margin in a two-dimensional feature space. The support vectors are bold, (a) for the separable case, and (b) for the non-separable case.

problem results in

$$L_P = \frac{1}{2}||\mathbf{w}||^2 + C\sum_i \xi_i - \sum_i \alpha_i\{y_i(\mathbf{x}_i \cdot \mathbf{w} + b) - 1 + \xi_i\} - \sum_i \mu_i\xi_i \qquad (3.36)$$

where the μ_i were introduced to enforce $\xi_i \geq 0$. The formulation of the dual problem L_D is the same as for the separable case (3.32). The only difference is an additional constraint $0 \leq \alpha_i \leq C \; \forall \; i$ meaning that in the non-separable case there exists an upper bound C on the α_i. For the computation of b again the KKT conditions of the primal problem are used, in particular, the complementary conditions

$$\alpha_i\{y_i(\mathbf{x}_i \cdot \mathbf{w} + b) - 1 + \xi_i\} = 0 \qquad (3.37)$$

and

$$\mu_i\xi_i = 0. \qquad (3.38)$$

Non-linear support vector machines

For data with a non-linear decision function the above described methods can be generalized. To this end, the training data is mapped to a higher dimensional (possibly infinite dimensional) Euclidean space $\mathcal{H}$ in which the data is linearly separable, using a mapping function Φ. Recall that for the training of the SVM in (3.32) as well as for the actual classification using (3.34) the data appears always in form of dot products $\mathbf{x}_i \cdot \mathbf{x}_j$. Thus, for data transformed to $\mathcal{H}$ the corresponding operation would be of the form $\Phi(\mathbf{x}_i) \cdot \Phi(\mathbf{x}_j)$. Suppose there exists a kernel function K with the property $K(\mathbf{x}_i, \mathbf{x}_j) = \Phi(\mathbf{x}_i) \cdot \Phi(\mathbf{x}_j)$. Next, we replace the dot product $\Phi(\mathbf{x}_i) \cdot \Phi(\mathbf{x}_j)$ in (3.32) and (3.34) by K which removes the mapping function Φ from the equations. Thus, we can avoid the actual mapping of the data to $\mathcal{H}$ and, moreover, we do not even need to explicitly know the mapping function Φ. Using

$$L_D = \sum_i \alpha_i - \frac{1}{2} \sum_{i,j} \alpha_i \alpha_j y_i y_j K(\mathbf{x}_i, \mathbf{x}_j) \tag{3.39}$$

and

$$f(\mathbf{x}) = \sum_i \alpha_i y_i K(\mathbf{s}_i, \mathbf{x}) + b \tag{3.40}$$

instead of (3.32) and (3.34) yields an SVM which performs a linear separation of the data in $\mathcal{H}$ corresponding to a non-linear separation in the lower dimensional original feature space. Several kernel functions with the property $K(\mathbf{x}_i, \mathbf{x}_j) = \Phi(\mathbf{x}_i) \cdot \Phi(\mathbf{x}_j)$ exist and have been studied for pattern recognition. A function can be used as kernel function if *Mercer*'s condition (see, e.g., Vapnik [172], Burges [23]) is satisfied. Mercer's condition requires that for any function $h(\mathbf{x})$ with

$$\int h(\mathbf{x})^2 d\mathbf{x} \quad \text{is finite} \tag{3.41}$$

then

$$\int K(\mathbf{x}, \mathbf{y}) h(\mathbf{x}) h(\mathbf{y}) d\mathbf{x} d\mathbf{y} \geq 0. \tag{3.42}$$

This condition checks whether a prospective kernel is a dot product in some space, but does not provide information on the mapping function Φ or the target space $\mathcal{H}$. Commonly applied nonlinear kernel functions are, e.g., the polynomial kernel of degree p

$$K(\mathbf{x}, \mathbf{y}) = (\mathbf{x} \cdot \mathbf{y} + 1)^p \tag{3.43}$$

or the Gaussian radial basis function (RBF) kernel

$$K(\mathbf{x}, \mathbf{y}) = \exp(-\gamma \|\mathbf{x} - \mathbf{y}\|^2) \quad \text{with } \gamma = \frac{1}{2\sigma^2}. \tag{3.44}$$

In our work we used the Gaussian radial basis function kernel since it proved to work very well for image classification problems in earlier work (e.g., Conrad *et al.* [41]).

Weighted support vector machines

Real world classification problems often provide unbalanced data sets, that is, the number of available data samples varies significantly between the classes. As a consequence, the classification error for the rarer classes usually is higher than for the more frequent classes. To limit the influence of the frequent classes in favor of the rare classes, but also to allow systematic weighting of classes, e.g., if one error type is more undesirable than another, the original formulation of SVMs has been extended to *weighted* SVMs (Osuna *et al.* [139]).

In (3.36) the penalty parameter C was introduced to realize a soft-margin classifier. Instead of using the same C for all classes as it was done there, weighted SVMs use class-specific penalty parameters C^+ and C^-. The previous objective function to be minimized

$$\frac{1}{2}||\mathbf{w}||^2 + C \sum_i \xi_i \qquad (3.45)$$

then becomes

$$\frac{1}{2}||\mathbf{w}||^2 + C^+ \left(\sum_{i:y_i=1} \xi_i \right) + C^- \left(\sum_{i:y_i=-1} \xi_i \right) \qquad (3.46)$$

while the minimization constraints (3.35) remain unchanged.

Multi-class classification

As described above, SVMs are designed for binary, i.e. two-class classification problems. Several methods have been proposed to extend SVMs for multi-class classification problems. Typically, a multi-class classifier is constructed by combining several binary classifiers. Common approaches to apply SVMs to multi-class problems are the *one-against-all* (Bottou *et al.* [19]), the *one-against-one* (Knerr *et al.* [102], Kressel [104]), and the *directed acyclic graph (DAG)* SVM (Platt *et al.* [151]). Also approaches that solve the multi-class problem using a multi-class SVM formulation have been described [173, 183]. To realize a k-class SVM, the objective function (3.45) used for a two-class SVM is generalized (Weston *et al.* [183]) to

$$\frac{1}{2} \sum_{m=1}^{k} ||\mathbf{w}_m||^2 + C \sum_i \sum_{m \neq y_i} \xi_i^m \qquad (3.47)$$

with minimization constraints

$$\mathbf{x}_i \cdot \mathbf{w}_{y_i} + b_{y_i} \geq \mathbf{x}_i \cdot \mathbf{w}_m + b_m + 2 - \xi_i^m \qquad (3.48)$$
$$\xi_i^m \geq 0, \quad i = 1, \dots, l \quad m \in \{1, \dots, k\} \setminus y_i.$$

However, this results in a more complex optimization problem and consequently, is computationally very demanding [88]. In Hsu *et al.* [88] the above listed approaches for multi-class classification one-against-all, one-against-one, DAG, and k-class SVM were compared experimentally, with the result that one-against-one and DAG performed best w.r.t. training time and classification accuracy.

The training algorithm for the one-against-one and the DAG method is the same. For a k-class problem $k(k-1)/2$ binary classifiers are constructed where each classifier is trained to separate two classes i and j. In the classification phase, the one-against-one approach uses a voting strategy. A data sample is classified by all binary classifiers and the class which results most frequently is chosen as overall output class (*max wins* strategy). The DAG approach uses a rooted binary directed acyclic graph in the classification phase. In this graph the $k(k-1)/2$ binary classifiers represent the internal nodes and the k classes the leave nodes. The classification process starts at the root node and the following node is chosen depending on the classification result. In each classification step one class is discarded and hence, when a leaf node is reached, only one class remains as the output class.

In this work the software library LIBSVM [27] has been used for SVM classification. This library implements the one-against-one approach for multi-class classification. The overall good performance of this approach was shown in [88].

3.6.3   Parameter optimization and cross validation

For SVMs with a Gaussian radial basis function kernel (3.44) there are two parameters which have to be selected: the error penalty parameter C and the kernel parameter γ which regulates the variance of the Gaussian kernel. To optimize C and γ we perform a grid search on a limited parameter space as proposed in Hsu *et al.* [87]. Therefore, different pairs (C, γ) are systematically created and used to train different SVMs. Their classification accuracy is compared and the pair (C, γ) which leads to the best result is selected as the optimal parameter set. The parameter space is sampled to define pairs (C, γ) using exponentially growing values of C and γ (e.g., $C = 2^{-5}, 2^{-3}, \ldots, 2^{15}$, $\gamma = 2^{-15}, 2^{-13}, \ldots, 2^3$). To determine the classification accuracy, one part of the training data is used for training of the classifier (training set), while the other part is used for testing of the trained classifier (test set). The percentage of correct classifications can be determined by comparing the classifier output with the true classes for the test set. Usually, confusion matrices are used to conveniently represent the classification accuracies and confusion probabilities (i.e. differences between true classes and classifier output, see Tab. 3.2).

However, training a classifier on one data set and testing it on another data set does not necessarily reflect the true classification accuracy since the performance naturally depends on the constellation of training and test set. Furthermore, if the parameters are selected based on only one training set they may be overadapted to

True	Classifier Output				
Class	1	2	3	$\cdots$	n
1	$TP1$	$FN1, FP2$	$FN1, FP3$	$\cdots$	$FN1, FPn$
2	$FN2, FP1$	$TP2$	$FN2, FP3$	$\cdots$	$FN2, FPn$
3	$FN3, FP1$	$FN3, FP2$	$TP3$	$\cdots$	$FN3, FPn$
$\cdots$	$\cdots$	$\cdots$	$\cdots$	$\cdots$	$\cdots$
n	$FNn, FP1$	$FNn, FP2$	$FNn, FP3$	$\cdots$	TPn

Table 3.2: Example confusion matrix for classes 1 to n. The matrix rows provide the true class, the columns provide the classifier output. *TP* are the true positives (i.e. correct classifications) which are given in the matrix diagonal. *FN* and *FP* are the false negatives and the false positives, respectively for each class.

the specific training set and the resulting SVM may perform poorly on other data sets (*overfitting* of the classifier). To achieve a more general and reliable result, the training and testing procedure for a parameter constellation (C, γ) can be repeated on several independent data sets using *cross validation*. For v-fold cross validation the training data is split into v equally sized subsets. Successively, $v - 1$ subsets are concatenated and used for training of the classifier which then is tested on the one remaining subset. This is repeated v times, until each subset has been used once for testing (*leave-one-out* strategy). Having determined the best combination (C, γ), the whole training set can be used again to train the final classifier. An example for parameter optimization using three-fold cross validation is shown in Fig. 3.18.

Furthermore, cross validation is an efficient technique to virtually increase the size of the test set since the whole training data set can be used for testing. Thus, cross validation can also be applied to determine a more general estimate of the overall performance of the classifier by using two nested cross validation loops. An example is shown in Fig. 3.19 where a ten-fold, outer cross validation is combined with a three-fold, inner cross validation (ten-times-three-fold cross validation).

3.7   Processing of Cell Cycle Phase Sequences

3.7.1   Introduction

After tracking and classification of the cell nuclei for all frames of an image sequence we obtain for each nucleus a sequence of cell cycle phases. Based on this result the goal of the final analysis step is to determine the duration of the cell cycle phases automatically. However, we cannot assume that the phase sequences are completely correct because of potential classification errors (for details on the classifier performance see Chapt. 4). Consequently, a suitable method for phase

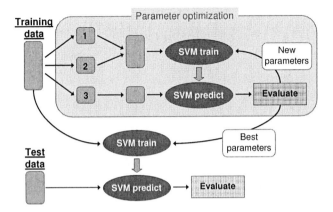

Figure 3.18: Three-fold cross validation for parameter optimization. The initial training set is split into three subsets and in each cross validation loop two subsets are used for training and one for testing. Finally, the optimal parameters and the complete training set are used to train the final classifier.

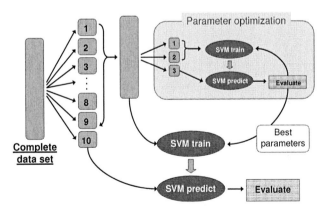

Figure 3.19: Two nested cross validation loops. Outer loop (ten-fold) to virtually increase the test set size, inner loop (three-fold) for parameter optimization.

length determination has to be able to deal with such errors. Here, we make use of prior knowledge on cell cycle progression, in particular, concerning biologically possible phase transitions. This allows us to build a model of the cell cycle which enables to check and recover the consistency of the phase sequences and determine the duration of the consecutive phases. In this thesis we developed a cell cycle model based on a finite state machine.

Finite state machines (FSMs) have been a topic in theoretical computer science since several decades (see, e.g., Booth [18], Minsky [127], Hopcroft *et al.* [86], Perrin [147]). In practical computer science they play an important role for compiler construction (Aho *et al.* [8]) and parsing of context-free grammars (Woods [186]). Furthermore, FSMs have been applied in a wide range of other application fields. One major application field is computer linguistics, where FSMs have been used, e.g., for natural language and speech processing (e.g., Roche *et al.* [157], Mohri [128]). But also in the image analysis field FSMs have been applied, e.g., for movie scene classification (Zhai *et al.* [196, 197]) or gesture recognition (Hong *et al.* [85], Yeasin *et al.* [193]).

Finite state machines are also closely related to the widely used hidden Markov models (HMMs). In essence, HMMs are FSMs which include probability density functions for all states, as well as transition probabilities for all transitions (see, e.g., Theodoridis *et al.* [169]). Consequently, HMMs are likewise applied, e.g., for speech recognition or image sequence processing. In Wang *et al.* [180], for example, a context-based mixture model (CBMM) was used to classify cell nuclei into four different cell cycle phases (interphase, prophase, metaphase, and anaphase). The CBMM presented there in essence is an HMM which considers not only previous, but also subsequent time steps for classification. We use an FSM for phase sequence analysis instead of an HMM, as for our type of analysis the FSM has several advantages. The main advantages are, first, we can directly formulate the allowed transitions in form of a transition table and we do not have to annotate a large amount of training sequences which include samples for all transitions. In Wang *et al.* [180], 100 single cell sequences were annotated manually for training, each including samples and transitions of the four considered classes. In our case, we are dealing with a much higher number of classes (up to 13), and often there is more than one possible transition. Consequently, we would require a much higher number of sequences for adequate training, which would mean a tremendous effort for manual annotation. Second, our FSM model is not adapted to the temporal characteristics of the training data (e.g., normal phase durations) in comparison to an HMM. If an HMM is trained, e.g., based on control sequences, it learns phase transition probabilities that are adapted to normal phase durations. However, for classification of phase sequences with prolonged or shortened phases (i.e. temporal phenotypes) this HMM most probably would not be accurate. Our FSM model, on the other hand, can likewise be applied to determine phase durations in control data as well as in data with temporal phenotypes. For a brief discussion of the applicability of HMMs in our application see also Sect. 5.2.

3.7.2 A finite state machine as cell cycle model

The biologically determined progression of phases during the cell cycle (see, e.g., Alberts *et al.* [9]) can be interpreted as a language which can be parsed using a finite state machine (FSM), in particular, an acceptor machine. An acceptor machine evaluates for a given sequence whether it is accepted by the machine or not, and generates a binary output accordingly. A cell cycle-related *phase sequence parser* has to implement all possible phases as its states and all possible phase transitions as respective directed state relations.

Formally, an acceptor machine is defined by a set of states $S = [s_1, \ldots, s_n]$, a set of input symbols $I = [i_1, \ldots, i_m]$ and a transition table $\delta : S \times I$ which provides the transfer function as a set of transition rules. In addition, all possible initial states S_i and final states S_f are specified. To check whether a given sequence of input symbols $i_t \in I$ is accepted by the FSM defined by (S, I, δ, S_i, S_f), the input is iteratively processed using the mapping

$$s_{t+1} = \delta(s_t, i_t). \tag{3.49}$$

We developed the FSM displayed in Fig. 3.20 which is suited to process sequences of seven different normal cell cycle phases (*interphase, prophase, prometaphase, metaphase, early anaphase, late anaphase*, and *telophase*). The set of states S includes for each phase one phase state as well as one error state, resulting in the total number of states $n = 14$. The input symbols I contain all mitotic phases represented by the integers $[1, .., 7]$. The FSM can be entered at any state except the error states, thus S_i contains the seven phase states, and it can be terminated at any state, i.e. $S_f \equiv S$. The transition table δ holds all biologically possible phase transitions as well as the error state transitions required for error correction (see Tab. 3.3). In practice, at each step of the phase sequence processing, one input phase is passed to the current active state of the FSM which is marked by a state pointer. If a transition to the state corresponding to the current input phase is possible, the state pointer moves to this particular state which becomes the new

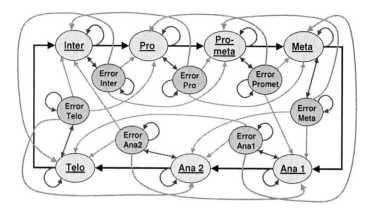

Figure 3.20: Finite state machine that accepts biologically consistent sequences of cell cycle phases. The states represent phases, the directed state relations represent the possible phase transitions (black arrows). Error states enable correction of consistency errors and therefore have relations to multiple phase states (gray arrows). For reasons of clarity only the most important relations are displayed.

active state. If the transition is not possible, i.e. an illegal transition in the input sequence occurred, the state pointer moves to the error state of the current active state. An error state is quit as soon as the current input phase is consistent with the previous sequence again. To prevent the FSM of getting stuck in an error state, the user can set a threshold on the number of steps the FSM can stay in an error state at maximum. If this threshold is reached the state pointer is reset, i.e. a new active state is selected according to the current input. Sequences which caused a reset of the state pointer can be finally discarded completely or partly.

To enable determination of phase durations each state implements a counter which logs the number of subsequent input phases while it is active. Once the state becomes inactive the respective phase identifier (input symbols $[1, .., 7]$) together with the measured phase duration is stored in a central phase lengths list. The error states use the symbol zero as phase identifiers to mark the processed phases as errors. When an error state is left, the error correction scheme is initiated (see Sect. 3.7.3 below). The final output of the FSM are the corrected input phase sequences as well as the overall phase length list for all processed phase sequences.

I	S						
	Inter	**Pro**	**Prometa**	**Meta**	**Ana 1**	**Ana 2**	**Telo**
1	Inter	E2	E3	E4	E5	E6	Inter
2	Pro	Pro	E3	E4	E5	E6	E7
3	E1	Prometa	Prometa	E4	E5	E6	E7
4	E1	E2	Meta	Meta	E5	E6	E7
5	E1	E2	E3	Ana 1	Ana 1	E6	E7
6	E1	E2	E3	Ana 2	Ana 2	Ana 2	E7
7	E1	E2	E3	E4	E5	Telo	Telo

I	S						
	E1	**E2**	**E3**	**E4**	**E5**	**E6**	**E7**
1	Inter	(Inter)	E3	E4	E5	Inter	Inter
2	Pro	Pro	(Pro)	E4	E5	E6	Pro
3	Prometa	Prometa	Prometa	(Prometa)	E5	E6	E7
4	E1	Meta	Meta	Meta	(Meta)	(Meta)	E7
5	E1	E2	Ana 1	Ana 1	Ana 1	(Ana 1)	E7
6	E1	E2	Ana 2	Ana 2	Ana 2	Ana 2	(Ana 2)
7	(Telo)	E2	E3	Telo	Telo	Telo	Telo

Table 3.3: Transition table δ for the FSM to process sequences of seven cell cycle phases as shown in Fig. 3.20. The table provides the successor state s_{t+1} for state $s_t \in S$ (given as columns) and an input $i_t \in I$ (given as rows). E1-E7 represent the error states. The brackets denote conditional transitions, i.e. they are only used in specific cases which are identified by the error correction method (see Sect. 3.7.3), otherwise the FSM stays in the current error state. Note that in Fig. 3.20 not all transitions are displayed for reasons of clarity. Meta–Ana 2 is defined as a valid transition since the relatively short early anaphase (Ana 1) in some sequences is not visible.

3.7.3   Error correction

As described above, we implemented the phase sequence parser as an acceptor machine, i.e. only biologically possible phase sequences are accepted. However, our FSM includes error states which are not necessarily final states since they involve an error correction strategy. This enables phase length determination also for sequences with inconsistencies of a certain length (selected by the user). Note that the longer an error is, the higher is the uncertainty of the correction. This is because the error correction interpolates a consistent phase sequence based on previous and subsequent phases in the sequence. To enable the correction of consistency errors of different types and lengths, the error states have relations to phase states in forward and backward direction. The more consecutive phase states are involved, the more complex errors can be corrected. The FSM developed here considers up to two states in backwards and up to three states in forward direction for error correction.

To illustrate the transition rules for the error states, Tab. 3.4 (left) shows an exemplary section of an FSM containing five normal states and one error state. In Tab. 3.4 (right) different erroneous input sequences are given in the table rows. For all cases, the input symbol at t_0 $i_{t_0} = 2$, and consequently the active state s_{t_0} is *state 2*. At t_1 an illegal input causes an error in all cases, resulting in $s_{t_1} = Error2$ (x denotes any illegal input symbol). The error correction is initiated at t_2 with a legal input for *Error 2* which determines s_{t_2}. An illegal input at t_2 would lead to $s_{t_2} = s_{t_1}$. The last column of Tab. 3.4 provides the possible corrections for the identified errors. The letters **a-d** given in the first column indicate the state relations which are used to exit *Error 2* for the different error types (see also Tab. 3.4 (left)). For the correction of error types **a**, **b** and **c** the input at t_1 and t_2 is taken into account, while for the ambiguous error types **a***, **c*** and **d** additionally the input at t_{-1}, and for **d** even at t_{-2}, is considered. Note that case **d** corresponds to the conditional transitions described in Tab. 3.3. If for case **d** we have $i_{t_{-2}} \neq 1$, the error state is not left in t_2 and consequently, the error correction is not initiated.

To choose the most appropriate correction if there exist different alternatives for error correction (e.g., **a***, **b**, **c*** in Tab. 3.4), we use knowledge on the accuracy of the classifier that produced the input sequence. The classification accuracy of the classifier usually is studied before applying it to unseen data, e.g., using cross validation techniques on the training data (see Sect. 3.6.3). Comparing the classifier output with the real classes for a large data set gives an estimate of the confusion probabilities between classes (e.g., Tab. 3.5). If for an occurring error in a phase sequence the correct class or the actual position of the error is not clearly defined by the context, we use the confusion probabilities to determine the most likely

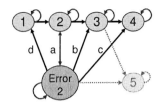

		Input				Possible
		t_{-2}	t_{-1}	t_0	t_1	t_2	corrections
a				2	x	2	222
a*			1	2	1	2	1112, 1222, 1122
b				2	x	3	223, 233
c				2	x	4	234
c*			2	2	4	4	2234, 2344
d		1	2	2	1	1	11111

Table 3.4: Schematic example for error correction. (Left) Exemplary FSM with phase states 1-5 and error state for phase state 2. All necessary state relations for *Error 2* are displayed: relation to predecessor of state 2, as well as to its first and second successors. If phase state 2 has multiple first and/or second successors (e.g., 4 and 5), the error state is related to all of them. (Right) Table of the correctable error types and the possible corrections. The first column (**a-d**) assigns each error type to the respective state relation. t_0 is the step directly before the error (active state: *2*), at t_1 the input causes an error (active state: *Error 2*), and at t_2 the input causes the exit of the error state (active state: depends on the correction). The x denotes any illegal input symbol. **a*** and **c*** are ambiguous errors, i.e. it is unclear whether the actual error occurs at t_0 or at t_1.

alternative. For the examples in Tab. 3.4 the correction of errors **a** and **c** is unique, in both cases there is only one reasonable solution. In case **b** the error is at a phase transition and the correct class could be either 2 or 3. Here, our algorithm checks whether it is more likely that a true class 2 is considered as x or that a true class 3 is considered as x (the x denotes the erroneous classifier output at t_1). That is

$$i_{t_1,correct} = \begin{cases} i_{t_0}, & if \quad P(i_{t_0}, i_{t_1}) \geq P(i_{t_2}, i_{t_1}) \\ i_{t_2}, & if \quad P(i_{t_0}, i_{t_1}) < P(i_{t_2}, i_{t_1}) \end{cases} \tag{3.50}$$

where P represents the confusion matrix and $P(a, b)$ is the confusion probability of the true class a and the classifier output b.

For errors **c*** and **a*** it is not clear whether i_{t_1} or i_{t_0} is the actual error. To find out which error is more likely for **c***, we check whether the lacking input symbol $i_{lack} = 3$ is more likely to be confused with 2 or with 4, i.e. whether $P(i_{lack}, i_{t_0}) \geq P(i_{lack}, i_{t_1})$ or $P(i_{lack}, i_{t_0}) < P(i_{lack}, i_{t_1})$. For **a*** it has to be determined whether it is more likely that 1 is confused with 2, or the other way round, i.e. whether $P(i_{t_0}, i_{t_1}) \geq P(i_{t_1}, i_{t_0})$ or $P(i_{t_0}, i_{t_1}) < P(i_{t_1}, i_{t_0})$. Using the example confusion matrix given in Tab. 3.5 results for **c*** in the corrected phase sequence 2234 and for **a*** in the corrected phase sequence 1112.

For errors that are longer than one time step our error correction approach does

True	Classifier Output			
Class	1	2	3	4
1	0.80	0.05	0.10	0.05
2	0.01	0.95	0.04	0.00
3	0.00	0.02	0.77	0.21
4	0.10	0.00	0.01	0.89

Table 3.5: Example confusion matrix providing the relative frequencies of confusion for classes 1-4 (corresponding to the error correction example in Tab. 3.4). The rows of the matrix give the true classes, while the columns show the classifier output. Hence, the diagonal provides the relative frequencies of true classifications.

not automatically correct the whole error sequence since this would require to trace back the error to its beginning. Recall that the error correction is performed at the end of an error, i.e. as soon as the error state receives a valid input symbol. Thus, only the time step directly before the valid input that leads to the initiation of the error correction is corrected. The other erroneous input symbols are replaced by zeros in the FSM output files, indicating an error. However, the FSM can be applied recursively on the corrected output sequences to successively correct longer errors. To interpolate a consistent phase sequence for an error of length k, the FSM has to be applied k times. As mentioned above, a maximum value for k can be specified by the user and the error state is terminated by a reset of the state pointer as soon as this value is reached. Consequently, if a reset occurs the sequence is no longer consistent and has to be discarded or postprocessed. In our application we used a value of $k = 2$.

3.7.4 Extended finite state machine for abnormal morphologies

The above described FSM is well suited to determine the cell cycle phase durations for normal cell cycle phases. To handle additional morphological phenotype classes occurring, e.g., when certain mitosis-related genes are silenced, we extended the FSM. The additional morphological phenotypes require new states and additional transition rules. The state set S for the extended FSM includes 12 phase states. In addition to the states mentioned above which represent the normal cell cycle phases we now include the following states representing morphological phenotypes: *abnormal interphase, abnormal early anaphase, abnormal late anaphase, abnormal telophase,* and *cell death.* Furthermore, we increase the number of error states to 12 accordingly, and I now contains the 12 allowed input symbols (integers $[1, .., 12]$).

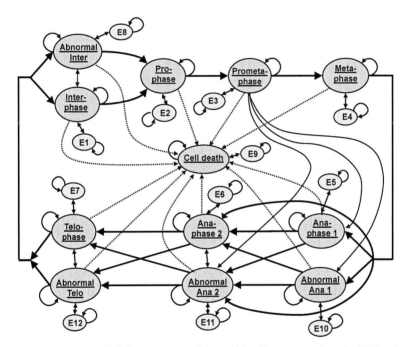

Figure 3.21: Extended finite state machine which allows processing of additional morphological phenotype classes. For reasons of clarity primarily the transitions of the phase states are displayed.

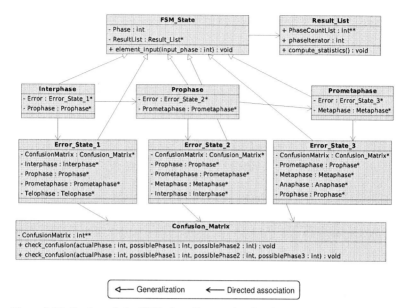

Figure 3.22: Section of the UML class diagram for the object oriented implementation of the FSM.

Figure 3.21 shows the extended FSM. Note that for reasons of clarity only the transitions of the phase states are displayed but not all transitions of the error states. With the additional phase transitions also the error state transitions have to be extended while the general functionality of the error correction is the same as described in Sect. 3.7.3 above.

However, the bidirectional relations between normal and abnormal phases (e.g., telophase – abnormal telophase) allow the FSM to accept not only biologically reasonable, incidental changes between the two states, but also less reasonable alternations between states (e.g., state changes at each time step). To improve such biologically implausible results we introduced an additional postprocessing procedure which identifies typical error sequences (like alternations) by searching for predefined patterns, and corrects the phase sequence according to biological constraints.

3.7.5   Implementation

For an object oriented implementation of the FSM we designed the class model shown in Fig. 3.22. Each state is represented by a state class which inherits from the state superclass `FSM_State`. The state relations are implemented as class attributes which are pointers on the respective state classes (e.g., `Prometaphase*`). The class `Result_List` is the central data structure that stores the determined phase durations of all states. Consequently, all state classes are related to `Result_List` to return the phase durations to the list at the end of their active phase. Error states additionally are able to modify the `Result_List` if previous results in the list have to be corrected. The class `Confusion_Matrix` provides the confusion probabilities of the input symbols which are required for the correction of errors with multiple correction alternatives. Since the error correction is regulated exclusively by the error states, `Confusion_Matrix` needs to be accessible only from the error states.

3.8 Data Visualization and Manual Annotation

3.8.1 Data representation

In this work, a large number of images with multiple objects per image frame were processed using a workflow which consists of different software components. Thus, data had to be represented in an exchangeable way that allowed transferring intermediate results as well as extracted information between subsequent processing steps in the image analysis pipeline. In our approach information extracted in the segmentation step is represented by the result images in which the objects of interest are segmented and labeled. In addition, text files are generated which list all objects of interest and contain for each object its label in the respective result image, the object's bounding box coordinates to enable fast access, and some basic features (e.g., size, mean intensity). However, for the representation of tracking results a list structure is less suitable since the trajectories of dividing cell nuclei have a tree-like structure. Therefore, tracking results are stored in XML (extensible markup language [3]) documents which enable a convenient representation of the tree-structured data. Each cell nucleus at each image frame is represented as a data object which basically contains the following items: (1) the current time step in the image sequence, (2) the coordinates of centroid and bounding box, (3) references on potentially available intermediate result images (e.g., segmented maximum intensity projection image or most informative slice), and (4) the class of the object which is provided by the automatic classifier or a human expert. The temporal correspondence of objects in consecutive image frames is represented by nesting of corresponding objects, i.e. with each time step the nesting level increases by one. Consequently, siblings within a trajectory always have the same nesting level.

To import XML files into an application, parsers are available for various programming languages (in our case we used the Apache Xerces parser which is available for C++, Java, and Perl [1]). Using the DOM (document object model) interface, the parser creates a tree structure within the respective programming language and allows iterating the tree using standard tree operations in the application. Edited XML files can also be exported using the DOM interface.

3.8.2   Visualization of the tracking and classification results

To visualize the tracking and classification results, a plugin has been developed for the public domain image processing software ImageJ [156]. ImageJ was chosen for the following reasons. First, this Java-based program can be easily extended by plugins and the developer can make use of the extensive application programming interface (API) ImageJ provides, containing routines, e.g., to load, display, and modify images and image sequences. Second, ImageJ is a well-known and widely accepted tool for image viewing and processing, in particular, among the target users in the biological application field. Third, ImageJ is platform-independent and thus, the software can be readily transferred between different operating systems without adaptations.

For visualization of the tracking and classification results we implemented the plugin *Track Class Overlay*. When the plugin is started the user can select the items to be visualized, i.e. the class information and/or the tracking information. The *Track Class Overlay* plugin uses the XML file containing the tracking and classification results as described above. While the user scrolls through the image stack a non-destructive overlay is generated on-the-fly at each image frame (i.e. no additional overlay images have to be stored or kept in the random access memory). Therefore, all available objects for this particular time step are collected from the data tree and the respective information is overlaid on the original image. Class information is represented by the bounding boxes where different classes are encoded in different colors. To visualize the tracking result, trajectories are displayed as lines connecting the gravity centers of corresponding objects. The maximum displayed length of the trajectories can be chosen by the user. This intuitive visualization allows convenient visual inspection of the generated results, in particular, for technically unversed users (see Fig. 3.23).

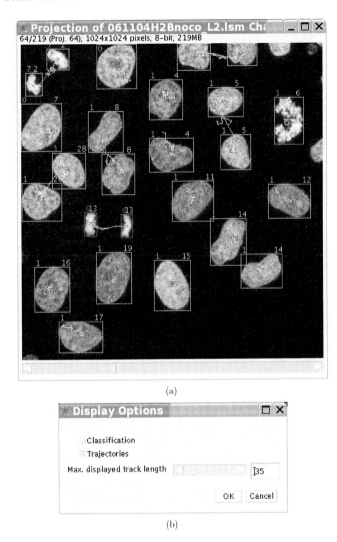

(a)

(b)

Figure 3.23: Screenshot of the ImageJ plugin *Track Class Overlay* that visualizes the trajectories and the assigned classes. (a) Window displaying the image sequence overlaid with the selected information. Here, the trajectories are displayed as yellow lines connecting the gravity centers of previous time steps (red dots). Also visible are the assigned cell cycle phases as colored boxes. At the upper left corner of a box the class is given as a number, at the upper right corner the cell number is displayed. (b) Dialog box to select the items to be displayed. For the trajectories the number of considered time steps can be selected.

3.8.3   Manual annotation of the training data

Our classification approach relies on supervised learning which allows accurate classification of cell nuclei. However, a supervised learning classifier requires ground truth data for training (see Sect. 3.6), and consequently the classifier can only be accurate if the training data is accurate. Moreover, the training data set has to be preferably large to ensure a good generalization performance of the classifier. Thus, the process of ground truth generation is an important step for the overall performance of the approach. In our case, the training set was generated by manual annotation of human experts. Since manual annotation of a large number of cell nuclei is very tedious and prone to errors, in particular, for multi-cell 3D image sequences, we developed a software tool to facilitate annotation. Again, this tool was realized as an ImageJ plugin which is based on the XML data file containing the tracking information.

The plugin *Image Annotation* traverses the XML tree and successively displays each cell nucleus for each time step with all available z-slices. The user can annotate the currently displayed object choosing from a list of predefined class labels. Also, the user can navigate in the trajectory tree by choosing *next* or *previous* object (where preorder tree traversal is used) or specifying cell number and time step to jump directly to one particular object (see Fig. 3.24). The assigned classes are stored in the XML file for each object. Finally, the manual annotation can be visually checked using the *Track Class Overlay* plugin as described above.

3.9   Summary

In this chapter we presented our approach for cell cycle analysis, consisting of approaches for segmentation, tracking, feature extraction, classification, and phase sequence parsing. For *segmentation* of cell nuclei we developed a region adaptive thresholding scheme that is based on overlapping image regions (see Sect. 3.3.3). After region adaptive thresholding, hole filling and labeling of the resulting objects is performed (see Sect. 3.3.4). We extended our segmentation approach to enable the segmentation of morphologically abnormal cell nuclei with detached chromosomes during mitosis or attached micronuclei (see Sect. 3.3.5). Our scheme for *tracking* of mitotic cell nuclei consists of two steps. First, one-to-one correspondences are established by searching for trajectories with maximum smoothness (see Sect. 3.4.2). Second, objects appearing at later time points are checked whether they result from a cell division. For detected mitosis events the respective trajectories are merged to establish one-to-many correspondences (see Sect. 3.4.3). We applied the overlap

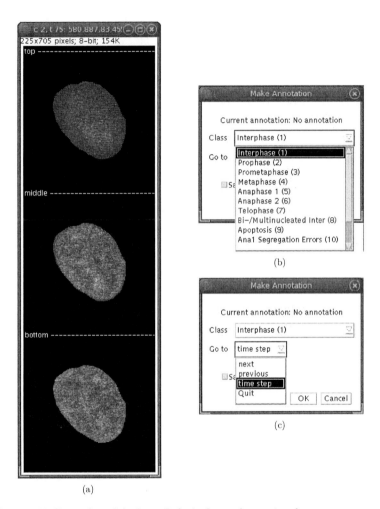

(a)

(b)

(c)

Figure 3.24: Screenshot of the ImageJ plugin *Image Annotation* that supports manual annotation for ground truth generation. (a) Window displaying the current cell nucleus with top, middle, and bottom slice. At the top of the frame the cell index, the current time step, and the centroid coordinates are displayed. (b) Annotation window with class choice. (c) Annotation window with navigation choice.

distance ratio for mitosis detection and also developed a new measure based on morphological criteria which significantly improved the mitosis detection results. For *feature extraction* based on 3D images we use the most informative slice as well as the maximum intensity projection image (see Sect. 3.5.1) because of the relatively low resolution in z-direction. We compute a large set of commonly used static image features as described in Sect. 3.5.2. In addition, we introduce dynamic image features which take into account the differences of basic features between corresponding cell nuclei in consecutive image frames (see Sect. 3.5.3). We perform feature normalization using the standard z-score method (see Sect. 3.5.4). Moreover, we applied different approaches for feature reduction based on feature recombination as well as on feature selection (see Sect. 3.5.4). For *classification* of cell nuclei we use support vector machines with a Gaussian radial basis function kernel. The adaptation to a multi-class classification problem is realized using a *one-against-one* approach (see Sect. 3.6.2). To optimize the parameters of the classifier C and γ, we perform a grid search on the parameter space using cross validation (see Sect. 3.6.3). Furthermore, cross validation is applied to determine a realistic estimate of the classification performance. We solved the problem of *phase length determination* using an error-correcting finite state machine. We developed an FSM that represents cell cycle progression, and enables to check and recover the consistency of phase sequences, and to determine the phase durations (see Sect. 3.7.2). For cases of ambiguous errors or correction we use the confusion probabilities of the classifier (see Sect. 3.7.3). Also, the FSM was extended to handle morphological phenotypes in addition to the normal cell cycle phases (see Sect. 3.7.4). For managing and exchanging the analysis results between the different software components we use XML. Based on this XML representation of the data we developed user-friendly tools for data visualization and annotation (Sect. 3.8).

Chapter 4

Experimental Results

4.1 Introduction

In this chapter, we describe the application of our approach to real fluorescence microscopy images and present experimental results. All of the data used here was acquired in the context of the EU project MitoCheck [5] by our cooperation partners (group of J. Ellenberg, European Molecular Biology Laboratory EMBL, Heidelberg). The overall goal of the MitoCheck project is to elucidate the coordination of mitotic processes in human cells at a molecular level, e.g., to provide a better understanding of the mechanisms of cancer development. To this end, high-throughput RNAi gene knockdown screens are performed to identify the genes involved in the process of mitosis based on a *primary screen*, and to study the functionality of the identified genes in more detail performing specific assays in *secondary screens*. Thereby, large amounts of live cell image sequences are acquired using automated microscopy. For a description of the methods for gene knockdown and automated microscopy for high-throughput screening see Sect. 1.2 above. In this work, we started with the analysis of images of the *pilot screen* which is part of the primary screen, and subsequently focused on the analysis of secondary screen image sequences which provide a higher information content and require more complex analysis methods.

The methods developed in this thesis have been used for the analysis of four different applications that are related to the MitoCheck project and will be described in the following sections. First, we present results for the pilot screen where the analysis is based on classification of cell nuclei into four phenotype classes using static 2D images (Sect. 4.2). For this task we apply segmentation, feature extraction, and classification. Second, results for image data from the secondary screen are presented. Here, the input data are 3D image sequences and the main analysis goal is to automatically determine the cell cycle phase durations. Therefore we apply our

Data	Applications			
Characteristics	§ 4.2	§ 4.3	§ 4.4	§ 4.5
Data dimensionality	2D	3D+t	3D+t	3D+t
Number of image sequences	-	4	48	951
Number of 2D images	40	1500	24,660	1,040,715
Size of single images [MB]	2.7	1	1	0.25
Number of z-slices	1	3	3	5
Total data volume [MB]	108	1500	24,660	∼250,000
Number of analyzed cells	637	10,416	157,771	2,229,482

Table 4.1: Characteristics of the four different applications our approach has been used on for analysis. The applications are described in this chapter in Sects. 4.2–4.5. Note that for the applications with data dimensionality 3D+t the second row 'Number of 2D images' provides the total number of image slices.

FSM-based approach for phase length determination after segmentation, tracking, and classification based on the 3D image sequences (Sect. 4.3). Third, we present results for a larger set of 3D image sequences with morphologically abnormal cell nuclei where we applied our extended methods for segmentation, tracking, and phase sequence parsing. Again, the task is to determine the mitotic phase durations and identify phase prolongations or shortenings induced by drug or siRNA treatment. For this application we not only determine the phase durations but also present a detailed statistical analysis comparing treated experiments with control experiments (Sect. 4.4). Fourth, we present results of the analysis of a complete high-content screen of nuclear envelope proteins (addressing 104 genes). There, the goal is to study the evolution of complex nuclear phenotypes over time. We analyzed the corresponding large amount of 3D image data (ca. 250 GB) using our extended approaches for segmentation, tracking, and classification (Sect. 4.5).

The four applications described above provide highly different amounts of data, from several 2D images (Sect. 4.2) up to a full screen including nearly one thousand 3D image sequences (Sect. 4.5). Table 4.1 gives an overview of the number of analyzed images together with the total data volume processed for each of the applications. Note that all described analysis tasks, in principle, could be performed on arbitrarily large data sets since the analysis workflow can be run fully automatically.

In all four applications the image acquisition was performed using automated fluorescence microscopy. With automated microscopy many different experiments can be imaged in parallel which is essential for high-throughput screens (see Sect. 1.2.4 above). For acquisition of the primary screen images, widefield fluorescence microscopy has been used since there the focus is on throughput rather than on resolution. For all other experiments a confocal fluorescence microscope has been used which provides images of higher resolution and in addition allows 3D imaging (see Sect. 1.2.2). All experiments are based on live cell microscopy and thus allow observing the temporal development of phenotypes. Chromosome morphology is visualized using a HeLa (Kyoto) cell line stably expressing the fluorescent chromosomal marker histone 2B-EGFP.

4.2 Classification of Cell Nuclei in 2D Images

4.2.1 Primary screen data

The aim of the MitoCheck primary screen is to identify the genes involved in mitosis. Therefore, a genome-wide RNAi screen is performed by systematically knocking down each single known human gene (approximately 22,000 protein coding genes). Image sequences are acquired for each knockdown experiment over an observation time frame of 44 hours using automated widefield fluorescence microscopy. At each spot (representing one knockdown experiment) an image is taken every 30 minutes which provides a relatively coarse temporal resolution since the mitotic process usually takes approximately one hour for HeLa cells. Thus, these images are less suited for single cell-based temporal analysis of the mitotic phases as the imaging frequency is not high enough to capture more than two images for a mitosis cycle. However, here the focus is on population-based temporal analysis and thus, the images are analyzed separately without considering the temporal context. In each image, cell nuclei are segmented and classified into four different phenotype classes (interphase, mitotic, apoptotic, and multi-nucleated cells). Finally, the percentage of each class can be analyzed throughout the observation time and compared between treated and control experiments. Significant abberations between treated and control experiments, such as, e.g., an increasing percentage of mitotic or apoptotic cells, indicate that the respective treatment has an influence on mitosis. Using this technique, a hit list of genes which affect mitosis when they are knocked down can be determined from the complete set of considered genes.

A pilot screen for the primary screen targeting 49 genes to validate the methodology has proven the applicability of this screening approach [136]. The work presented here is based on images of this pilot screen which has been performed using the same experimental settings as for the complete primary screen. All images have a gray value depth of 12 bit and a resolution of 1344×1024 pixels ($10\times$ objective). A typical image contains about 200–400 nuclei with an average diameter of approximately 30 pixels (see Fig. 4.1).

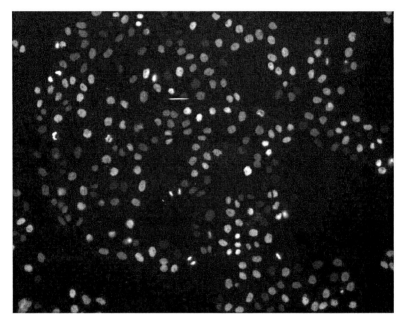

Figure 4.1: Typical example image of the primary screen.

Algorithm	Global	Strategy (1) (application to central pixel)	Strategy (2) (application to whole region)
Time [sec]	0.0875	60.1281	0.1812
Accuracy [%]	55.9	98.0	92.1

Table 4.2: Average time consumption in seconds on an AMD Opteron processor (1.8 GHz) for one image (1344 × 1024 pixels × 12 bits) and segmentation accuracies for the different thresholding techniques.

Classes	Training set	Test set	Total
Interphase	258	64	322
Mitosis	88	22	110
Apoptosis	48	12	60
Shape	116	29	145
Total	510	127	637

Table 4.3: Number of single cell image samples per class, and composition of the training set and test set.

4.2.2 Approach

To classify cell nuclei into phenotype classes based on 2D multi-cell images we applied the workflow described in Sect. 3.2.1, including segmentation, feature extraction, and classification. We determined the segmentation accuracy based on four typical images for global thresholding and two region adaptive thresholding strategies (strategy (1) and (2) as described in Sect. 3.3.3 above). As shown in Tab. 4.2 global thresholding was fastest but yielded the lowest accuracy of 55.9%. Since strategy (2) was significantly faster than strategy (1) (factor of 332), and still yielded an acceptable accuracy of 92.1%, we used strategy (2) in this application which applies the threshold to the whole region. To validate our approach, a representative set of single cell nuclei from 40 multi-cell images (taken from different experiments) was manually classified by experts in cell biology at EMBL. The objects were assigned to four classes: (1) *interphase*, (2) *mitosis*, (3) *apoptosis* (cell death phenotypes), and (4) *shape* (clustered nuclei). Example images for all classes are given in Fig. 4.2. The total number of manually classified cells is 637 and the number of cells per class is provided in Tab. 4.3 (note that from each multi-cell image only around 10-20 cell nuclei were selected). After segmentation of the respective 40 multi-cell images we computed a set of 353 features for each cell nucleus. The number of extracted features for each feature group is given in Tab. 4.4. Finally, the extracted features were used for supervised classification of cell nuclei.

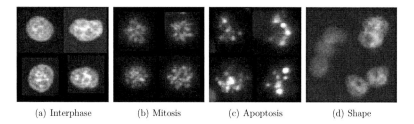

(a) Interphase   (b) Mitosis   (c) Apoptosis   (d) Shape

Figure 4.2: Example images representing the four classes considered in the primary screen data. Class apoptosis (c) includes cell death phenotypes, and shape (d) represents bi- and multi-nucleated cells. The example nuclei were cut out of the original images and artificially combined into the four images shown here.

Feature type	Number
Basic object-/edge-related	11
Granularity	21
Tree-structured wavelets	2
Haralick texture	260
Gray scale invariants	10
Zernike moments	49

Table 4.4: Different types of features used for classification and number of features computed for each type.

4.2.3   Experimental results

In our approach we use support vector machines (SVMs) for classification. In addition, we have also analyzed the performance of other classifiers in combination with feature reduction techniques. Below, we first describe the analysis results using our approach based on SVMs, and second, present the results of experiments with five other classifiers and SVMs.

Classification based on support vector machines

We validated our approach using the above described manually classified image data. Therefore, we split the available samples for each class randomly in training data and test data at the ratio of 4:1, resulting in a training set size of 510 and a test set size of 127 samples. Table 4.3 shows the number of training and testing samples per class. For the training set we standardized each feature to a mean value of zero and unit variance as described in Sect. 3.5.4. In the test set, the feature values were linearly transformed based on the transformation parameters from the training set. Then we trained an SVM classifier with a radial basis function (RBF) kernel (see Sect. 3.6.2) based on the training data set as described above and subsequently applied the SVM to classify the test set samples. An evaluation of the result yielded an overall classification accuracy of 96.9%. Thereby, 123 out of the 127 test set samples were correctly classified. Since here the number of test samples was high enough given the number of classes, the obtained result already provides a sound estimation of the classification accuracy. However, to check the reliability of this result we repeated the classification step, applying a ten-fold outer cross-validation on the whole data set of 637 images. This classification yielded an overall accuracy of 96.1%. Thus, both classification results correspond very well and we can conclude that an overall classification accuracy of around 96% is a reliable estimate.

The confusion matrices of both classification experiments (see Tabs. 4.5 and 4.6) reveal that misclassifications mostly occurred between the classes *mitosis* and *apoptosis*. Such misclassifications are in many cases caused by the similarity between samples of both classes. This is illustrated by the two examples in Fig. 4.3 showing cell nuclei in *mitosis* and *apoptosis* respectively, which even a human observer hardly can assign to either class based on a single image. Note that for manual image annotation, e.g., for ground truth generation, previous and subsequent image frames are considered to determine the true class in such cases. Moreover, the lower classification accuracy of the *apoptosis* class is also caused by the comparatively low number of 60 samples for this class (Tab. 4.3).

True	Classifier Output				Accur.
Class	Interph.	Mitosis	Apoptos.	Shape	[%]
Interphase	**64**	0	0	0	**100**
Mitosis	0	**20**	2	0	**91**
Apoptosis	0	2	**10**	0	**83**
Shape	0	0	0	**29**	**100**

Table 4.5: Confusion matrix for SVM classification. Here, the classifier was trained using 510 samples and tested on 127 samples. The composition of training and test set is given in Tab. 4.3. The overall accuracy is 96.9%.

True	Classifier Output				Accur.
Class	Interph.	Mitosis	Apoptos.	Shape	[%]
Interphase	**320**	1	0	1	**99.38**
Mitosis	0	**100**	10	0	**90.91**
Apoptosis	1	12	**47**	0	**78.33**
Shape	0	0	0	**145**	**100**

Table 4.6: Confusion matrix for SVM classification. Here, ten-fold cross validation was used on the complete annotated data set (637 samples, see Tab. 4.3) to virtually increase the test set. The overall accuracy is 96.1%.

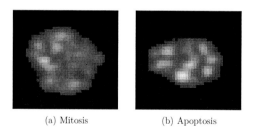

(a) Mitosis   (b) Apoptosis

Figure 4.3: Example for a pair of samples of the classes mitosis and apoptosis which look very similar and were confused by the classifier. (a) and (b) provide the true class annotation.

Comparison of different classifiers using feature reduction

Here, the primary objective was to investigate different methods for efficient feature space reduction and to compare different classification methods. The same image set and basically the same features as described above were used. First, different classifiers were applied based on the original feature set, and second, the original feature space was reduced and the classification experiments were repeated [103]. In this study, six different classification methods were examined including k-means clustering, hard competitive learning, and the neural gas (soft competitive learning) method [48], as well as hierarchical clustering [21, 170], support vector machines [49], and random forests [20, 114]. Note that the first three methods are unsupervised clustering methods and the last three methods supervised learning methods. For feature space reduction we applied principal component analysis (PCA) and independent component analysis (ICA) (see Sect. 3.5.4). All classification experiments have been performed using the statistical computing environment R [154].

To determine the classification accuracies for the different classifiers, the following test scheme was repeated 500 times in order to obtain reliable estimates. At each test loop the input dataset consisting of 636 cell feature vectors was randomly split (i.e., with uniform probability) into training and test sets consisting of 318 cells each. Accordingly, in each subset the number of cells of each class was proportional to the original number of cells available. Then both sets were applied for examining all classifiers so that exactly the same data was used for the compared classifiers. For each classification method the overall accuracy as well as a confusion matrix was computed. Finally, the mean accuracy values (with standard deviations) and mean confusion matrices were calculated. Figure 4.4 shows a sketch of this procedure.

The mean classification accuracies for all combinations of classifiers and feature reduction methods are given in Tab. 4.7. Among the six examined classifiers the lowest classification accuracies in the range of 55–60% resulted from the unsupervised clustering methods as it could be expected (note that random classification in case of four classes on the average has an accuracy of 25%). The hierarchical clustering method yielded medium accuracies in the range of 78–83%. The best results were achieved by SVMs and random forests with the highest accuracies of 94.37% and 94.03%, respectively. Even in combination with the different feature reduction strategies, these methods still produced classification accuracies in the range of 83–94% with similar standard deviations. Tabs. 4.8 and 4.9 provide the mean confusion matrices for the best combinations, i.e. SVMs using 10 principal components and random forests using all original features. Again, the confusion matrices reveal that misclassification occurred mostly between the classes mitosis

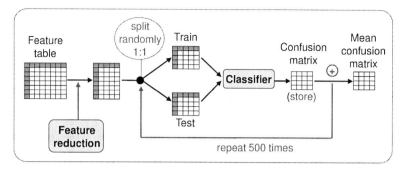

Figure 4.4: Test scheme used to compare different classifiers and feature reduction methods. In comparison to leave-one-out cross validation, here data samples can be used multiple times for testing. Only the composition of training and test set, which is determined randomly is different for each run.

and apoptosis which can be explained by morphological similarities of both classes as shown above (Fig. 4.3). Also we can observe that for the combination PCA and SVMs the classification error is less biased by the unbalanced training data, i.e. the accuracy for interphase (for which the most training samples are available) is reduced in favor of the rarer classes. Regarding the computation time we found that SVMs are significantly faster than random forests (128 sec compared to 1975 sec, using all original features for 500 training and test cycles on a Pentium 4, 3.2 GHz computer). Concerning the performance of the feature reduction strategies, PCA in all cases well preserved the classification accuracies (+/-2%), while ICA performed slightly worse (see Tab. 4.7). However, the application of feature reduction methods lowered the computation time significantly, e.g., using the 10 leading principal components resulted in a reduction of the computation time between a factor of 10.8 (random forests) and 33.9 (hard competitive learning).

4.2.4   Summary and conclusion

In this section we presented results of our approach applied to 2D multi-cell images. We showed that our analysis approach allows classifying cell nuclei into four classes (interphase, mitosis, apoptosis, shape) with an overall classification accuracy of above 96%. For the classes interphase and shape even an accuracy of almost 100% was reached. Thus, our scheme proved to be well suited for accurate, automatic evaluation of large image sets. In addition, we studied the performance of support vector machines in comparison to five other classifiers regarding classification accuracy and computation speed. It turned out that among the compared methods SVMs as well

Method	Mean accuracy (STD) [%]						
	358 FEAT	5 PCs	10 PCs	23 PCs	5 ICs	10 ICs	23 ICs
K-means	58.26 (4.58)	59.99 (3.21)	57.93 (4.60)	57.84 (4.81)	55.12 (6.76)	59.97 (7.76)	59.32 (7.53)
Hard Comp. Learning	57.05 (5.42)	60.21 (4.06)	58.37 (5.57)	58.51 (5.22)	55.21 (7.73)	58.28 (8.55)	57.55 (6.43)
Neural Gas	58.56 (4.58)	59.60 (3.01)	57.39 (4.07)	57.14 (4.21)	56.95 (5.19)	61.33 (7.74)	63.09 (7.77)
Hier. Clustering	82.94 (5.86)	81.30 (5.42)	80.84 (5.84)	81.85 (5.58)	77.93 (4.89)	78.36 (5.43)	78.13 (5.68)
SVMs	92.98 (2.37)	91.93 (1.48)	**94.37** (1.25)	92.87 (1.45)	83.22 (1.66)	90.77 (1.53)	92.96 (1.42)
Random Forests	**94.03** (1.33)	90.88 (1.34)	92.49 (1.31)	92.93 (1.26)	87.09 (1.72)	90.34 (1.56)	90.86 (1.52)

Table 4.7: Mean and standard deviation (in brackets) of cell classification accuracy for 6 classification methods when using 358 original features as well as 5, 10, and 23 principal components (PCs) covering 90, 95, and 99 percent of original feature variance, respectively, and using 5, 10, and 23 independent components (ICs). Accuracy statistics were computed over 500 replications. The two highest accuracies are printed in bold face.

True Class	Classifier Output				Accur. [%]
	Interph.	Mitosis	Apoptos.	Shape	
Interphase	**158.2**	0.4	0.0	1.5	**98.81**
Mitosis	0.8	**47.9**	5.4	0.8	**87.25**
Apoptosis	2.2	6.0	**21.3**	0.8	**70.30**
Shape	0.0	0.0	0.0	**72.8**	**100**

Table 4.8: Mean confusion matrix for SVM classifier using 10 leading principal components (computed over 500 replications).

True Class	Classifier Output				Accur. [%]
	Interph.	Mitosis	Apoptos.	Shape	
Interphase	**158.8**	0.6	0.4	0.8	**98.88**
Mitosis	0.3	**47.0**	7.8	0.0	**85.30**
Apoptosis	1.1	7.6	**20.8**	0.5	**69.33**
Shape	0.0	0.0	0.0	**72.4**	**100**

Table 4.9: Mean confusion matrix for random forests classifier using 358 original features (computed over 500 repetitions).

as random forests provide the best results. However, SVMs were significantly faster than random forests. This confirms our choice to use support vector machines in our overall approach. Also we showed that principal component analysis (PCA) is a well suited technique for feature space reduction in our application.

The results described in this section have been published in conference proceedings [81, 82, 103], as well as in a book chapter [73].

4.3 Determination of Mitotic Phase Durations Based on 3D Image Sequences

4.3.1 Secondary screen data

The goal of the MitoCheck secondary screen is to study the genes identified in the primary screen in more detail to gain information on the functionality of the respective genes. Based on the primary screen, several hundreds of genes have been identified which have a general influence on mitosis. For the secondary screen these genes are clustered in functional groups and specific assays are designed to study the particular properties of the gene groups. Consequently, such assays have specific experimental settings regarding, e.g., the number of imaging channels and the type of fluorescently labeled structures, or the spatial and temporal resolution. To study cell cycle progression in general, the chromatin structure provides the crucial information, however, also additional structures such as microtubles (which form the mitotic spindel), the nuclear membrane or kinetochores can be imaged to gain additional information. Here, we analyzed an assay to study mitotic delays, i.e. prolongations of single mitotic phases. In particular, the images analyzed in this section were acquired in RNAi knockdown experiments targeting the enzyme subunit PNUTS (protein phosphatase-1 nuclear targeting subunit). PNUTS has been shown to be important for chromosome decondensation in telophase (Landsverk et al. [107]) as well as for chromosome condensation in prophase (Mora-Bermúdez [129]). For more details on the PNUTS knockdown experiments see [129]. In this assay only the chromatin was visualized since the mitotic phases can be well identified based on the chromatin structure. To allow a fine subdivision into mitotic phases, here a significantly higher temporal and spatial resolution was used than in the primary screen. A confocal fluorescence microscope was used to acquire three image slices per time step at each location with a spatial resolution of 1024×1024 pixels ($63 \times$ objective) and a gray value depth of 8 bit. The image acquisition interval was six to seven minutes, and the total observation time was around 14 hours. One image

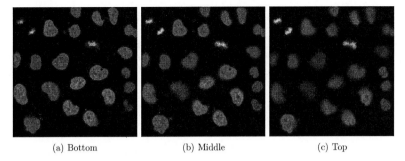

(a) Bottom	(b) Middle	(c) Top

Figure 4.5: Typical example image stack of the mitotic delay assay of the secondary screen. The three confocal slices for one time step are shown.

contains about 20 nuclei where nuclei in interphase have an average diameter of approximately 100-150 pixels (see Fig. 4.5).

4.3.2 Approach

To analyze the 3D image sequences acquired for the mitotic delay assay of the secondary screen we applied the workflow described in Sect. 3.2.2. We validated our approach based on four 3D multi-cell image sequences. For fast and robust analysis, segmentation and tracking were performed on maximum intensity projections of the original images. Applying maximum intensity projection for all stacks of all sequences resulted in 500 projected multi-cell images. For mitosis detection in the tracking step we here used the overlap distance ratio criterion since the new morphology-based criterion (see Sect. 3.4.3) was not yet developed. Using the overlap distance ratio, the tracking scheme was able to detect 80.0% of all occurring mitosis events (determined by visual inspection). Since the subsequent processing steps rely on correctly detected mitosis events we corrected the remaining ones manually. Static and dynamic features were computed based on the original image slices, in particular, for each cell nucleus the most informative slice was selected automatically at each time step. For most informative slice selection we investigated the maximum entropy as well as the maximum total intensity measure (see Sect. 3.5.1). Based on visual inspection of the selected slices we found that the maximum entropy measure was less suited for our application since it is relatively sensitive to noise (see Fig. 3.13). Consequently, we chose to use the maximum total intensity measure. To provide ground truth for the classification experiments, all contained cell nuclei were manually classified by experts into seven classes: (1) *interphase*, (2) *prophase*, (3) *prometaphase*, (4) *metaphase*, (5) *early anaphase* (*ana1*), (6) *late anaphase* (*ana2*),

Inter	Pro	Prometa	Meta	Ana1	Ana2	Telo
1143	*79*	*56*	*120*	*17*	*89*	*263*

Figure 4.6: Example images of the different mitotic phases and number of available samples per class.

and (7) *telophase* (see Fig. 4.6). Then, different classification experiments were performed as described in the following section. Finally, the phase sequence parser was applied to check and recover the consistency of the phase sequences, and to determine the phase durations.

4.3.3   Experimental results

In this application, we performed different classification experiments based on differently composed feature sets. In particular, we first studied the gain provided by the dynamic features, and second, we investigated three different feature selection strategies. For all classification experiments the feature values were normalized and a support vector machine classifier with a radial basis function kernel was used. Finally, we applied our phase sequence parser on 22 completely tracked and classified cell lineage trees (16 controls and 6 treated) and determined the phase durations.

Dynamic features

To examine the effect of including dynamic features, we performed classification on the whole annotated data set applying five-fold outer cross validation (for the numbers of samples per class see Fig. 4.6). Note that in this experiment the number of interphase samples was reduced by systematic sampling to 1143 samples (i.e. each nth interphase sample was picked from the sequence of interphase samples, see, e.g., [84]). In the first step, we used the whole set of 349 features including six dynamic features (differences to predecessor and successor for *object size, mean intensity,* and *standard deviation of intensity*). As shown in Tab. 4.11, we obtained classification accuracies of 64.7% to 97.5% for the different phases and an overall accuracy of 94.6%. In the second step, we only used the static features and repeated the classification experiment. In this case, we obtained a lower overall accuracy of 92.9% and single class accuracies in the range of 29.4% to 97.0% (see Tab. 4.10). Thus we can

| True | Classifier Output | | | | | | | Accur. |
Class	Inter	Pro	Prom.	Meta	Ana1	Ana2	Telo	[%]
Inter	**1109**	8	0	1	0	0	26	97.0
Pro	11	**66**	1	0	1	0	0	83.5
Prom.	0	0	**44**	6	6	0	0	78.6
Meta	1	0	5	**110**	0	0	4	91.7
Ana1	1	0	5	3	**5**	2	1	29.4
Ana2	0	0	0	0	1	**76**	12	85.4
Telo	21	0	0	1	2	8	**231**	87.8

Table 4.10: Confusion matrix and classification accuracies using five-fold cross validation; dynamic features were *not included*; overall accuracy: 92.9%.

| True | Classifier Output | | | | | | | Accur. |
Class	Inter	Pro	Prom.	Meta	Ana1	Ana2	Telo	[%]
Inter	**1114**	6	0	1	0	0	22	97.5
Pro	5	**72**	2	0	0	0	0	91.1
Prom.	0	0	**51**	5	0	0	0	91.1
Meta	0	0	2	**114**	1	1	2	95.0
Ana1	1	0	2	2	**11**	1	0	64.7
Ana2	0	0	0	0	1	**84**	4	94.4
Telo	33	0	0	1	1	3	**225**	85.6

Table 4.11: Same as Tab. 4.10 but *including* dynamic features; overall accuracy: 94.6%.

conclude that *including* dynamic features significantly improved our result. Comparing the accuracies for the single classes of both experiments as listed in Tabs. 4.10 and 4.11 shows that all class accuracies are significantly higher (except for *telophase*) if dynamic features are included. The largest improvement can be observed for *early anaphase* with an increase from 29.4% to 64.7% and for *prometaphase* with an increase from 78.6% to 91.1%. The generally lower accuracy for *early anaphase* is due to the relatively small number of available samples, in combination with the large morphological variability within this class (see Fig. 4.7). This unfavorable combination is caused by the fact that in *early anaphase* large structural changes of the chromatin happen in a time frame which is shorter than the sampling interval. Thus, not for all mitosis events an image of *early anaphase* is captured and the samples of this class span a wide morphological variability.

Feature selection experiments

We also investigated three different feature selection schemes and performed classification experiments based on training and test set without cross validation. There-

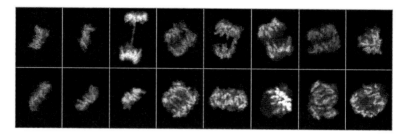

Figure 4.7: Training samples available for class *early anaphase*. In some cases both
daughter chromosome sets are still connected and thus treated as one object, in
other cases the chromosomes are already completely separated and consequently
only one daughter chromosome set is visible.

fore, we split the available samples for each class randomly into training data and
test data at a ratio of 2:1. Compared to the experiments in the previous paragraph,
the feature set here included six additional dynamic features, i.e. differences of shape
features, resulting in 355 features in total. Again, we used the complete feature set
for classification in the first step, which resulted in an overall classification accuracy
of 97.6% and per class accuracies as given in Tab. 4.12. Then we applied SAM
(significance analysis of microarrays) feature selection (Tusher *et al.* [171]) which
produces a ranked list of all features. We used the 50 top ranked features for classification based on exactly the same data set as before, which resulted in a slightly
decreased overall accuracy of 96.9%. As can be seen in Tab. 4.13, the accuracies for
prophase and *metaphase* were significantly decreased, however, *prometaphase* was
classified with an increased accuracy of 100%. Next, we used stepwise discriminant
analysis (Leray *et al.* [108]) to determine the best 128 features from the original
feature set. Based on this feature subset an overall classification accuracy of 97.6%
was yielded. Once again, the classification accuracy for *prophase* was decreased,
in this case in favor of *late anaphase* (see Tab. 4.14). Finally, we tested a feature
selection method based on the classification accuracies for single features (Puig *et*
al. [153]). The 150 top ranked features were used for classification resulting in an
overall classification accuracy of 96.3%. Using this method, the accuracy for early
anaphase was decreased strongly to 20%. Most of the other single class accuracies
were slightly decreased compared to classification based on the original feature set
(see Tab. 4.15).

    In conclusion, all feature selection strategies almost retained the overall classi-
fication accuracy compared to the results for the complete feature set. The third
method based on single feature classification performed somewhat worse, whereas

the SAM and stepwise methods provided good results. However, even though the latter two methods were able to increase single class accuracies to values up to 100%, the overall performance was not improved since other class accuracies were decreased at the same time. In particular, for the most challenging class *early anaphase* none of the studied feature selection methods was able to improve the classification performance.

Phase sequence analysis

To test the performance of the phase sequence parser we selected 29 cell lineage trees out of the four annotated image sequences including 4225 single cell 3D image stacks. For classification the whole feature set was used and two-fold cross validation was applied where the cross validation sets were created based on complete cell lineage trees. Because of the relatively low number of 29 cell lineage trees we did not split the data at a ratio of 1:1 but used 18 cell lineage trees for training and 11 trees for testing in each loop. Thus, in total only 22 trees out of the available 29 trees could be used for testing, however, using the unusual splitting ratio of 18:11 here was necessary to guarantee a sufficiently large sample number for each class in the training sets (the sample numbers for both training and test sets are given in Tab. 4.16). In each cross validation loop the classification accuracy and the confusion probabilities were determined based on the current training set using five-fold cross validation for later use in the phase sequence parser. Then, a classifier was trained with the current training set and tested with the respective test set. After the two runs we obtained an overall classification accuracy of 96.6% and the combined confusion matrix shown in Tab. 4.17. In agreement with previously described experiments the highest accuracy of 99.0% was reached for *interphase*, while the lowest accuracy of 57.1% resulted for *early anaphase*. The phase sequences obtained for the 22 classified cell lineage trees were processed with the phase sequence parser described in Sect. 3.7.2 using the classifier confusion matrices (which were determined during cross validation) for error correction. Finally, the corrected phase sequences were compared to the manually assigned true annotation (ground truth). It turned out that all inconsistencies with a duration of one time step were resolved in the first run of the finite state machine (FSM). Applying the FSM a second time on the corrected sequences additionally resolved all inconsistencies with a duration of two time steps. Longer inconsistencies were detected and partly resolved. In total 25 phases were corrected after the second run. However, errors which do not violate the phase sequence consistency such as, e.g., at the transition of two phases, are not detectable solely based on the phase sequence (neither automatically nor manually).

True	Classifier Output							Accur.
Class	Inter	Pro	Prom.	Meta	Ana1	Ana2	Telo	[%]
Inter	**1235**	1	0	0	1	0	3	99.6
Pro	5	**16**	1	0	0	0	0	72.7
Prom.	0	0	**13**	2	0	0	0	86.7
Meta	1	0	0	**32**	0	0	0	97.0
Ana1	0	0	0	2	**3**	0	0	60.0
Ana2	0	0	0	1	0	**21**	1	91.3
Telo	15	0	0	0	0	1	**52**	76.5

Table 4.12: Confusion matrix for classification based on the complete original feature set; overall accuracy: 97.6% (1372/1406).

True	Classifier Output							Accur.
Class	Inter	Pro	Prom.	Meta	Ana1	Ana2	Telo	[%]
Inter	**1232**	0	3	0	0	0	5	99.4
Pro	11	**11**	0	0	0	0	0	50.0
Prom.	0	0	**15**	0	0	0	0	100.0
Meta	2	0	2	**28**	0	0	1	84.9
Ana1	0	0	1	1	**3**	0	0	60.0
Ana2	0	0	0	1	0	**22**	0	95.7
Telo	14	0	0	1	0	1	**52**	76.5

Table 4.13: Confusion matrix for classification based on 50 features selected using SAM feature selection; overall accuracy: 96.9% (1363/1406).

True	Classifier Output							Accur.
Class	Inter	Pro	Prom.	Meta	Ana1	Ana2	Telo	[%]
Inter	**1237**	1	0	0	0	0	2	99.8
Pro	5	**15**	1	1	0	0	0	68.2
Prom.	0	0	**13**	2	0	0	0	86.7
Meta	0	0	0	**31**	0	0	2	93.9
Ana1	0	0	0	2	**3**	0	0	60.0
Ana2	0	0	0	0	0	**23**	0	100.0
Telo	16	0	0	0	0	2	**50**	73.5

Table 4.14: Confusion matrix for classification based on 128 features identified by stepwise discriminant analysis; overall accuracy: 97.6% (1372/1406).

True	Classifier Output							Accur.
Class	Inter	Pro	Prom.	Meta	Ana1	Ana2	Telo	[%]
Inter	**1234**	0	0	0	0	0	6	99.5
Pro	6	**15**	0	1	0	0	0	68.2
Prom.	0	0	**11**	3	1	0	0	73.3
Meta	0	2	3	**28**	0	0	0	84.9
Ana1	0	0	2	2	**1**	0	0	20.0
Ana2	0	0	0	1	0	**22**	0	95.7
Telo	21	0	0	0	0	4	**43**	63.2

Table 4.15: Confusion matrix for classification using 150 features selected using a classification accuracy-based method; overall accuracy: 96.3% (1354/1406).

Run 1	Inter	Pro	Prom.	Meta	Ana1	Ana2	Telo
Train	2200	43	28	59	9	46	121
Test	1520	25	17	42	7	23	85
Run 2	**Inter**	**Pro**	**Prom.**	**Meta**	**Ana1**	**Ana2**	**Telo**
Train	2254	42	26	67	9	42	139
Test	1466	26	19	34	7	27	67

Table 4.16: Number of training and testing samples per class for the classification experiment with complete cell lineage trees.

Unfortunately, classification errors at the transition of two phases occur with higher probability, as can be seen in the confusion matrix in Tab. 4.17. A quantitative evaluation of the frequency of detectable and undetectable errors for the here analyzed data showed that undetectable errors occur with approximately twice the frequency of detectable errors. Nevertheless, the correction of inconsistencies is particularly important to allow automatic determination of the phase durations.

Figure 4.8 shows an example for a resulting phase sequence where all phases were computed correctly, e.g., for *prophase* the computed length is two time steps. We determined all phase durations for the 22 classified cell lineage trees. The data consisted of 16 cell lineage trees from three different control experiments, and 6 cell lineage trees from one treated experiment. Figure 4.9 shows the resulting mean phase durations for the treated and control samples. For *interphase* the data is not shown since this phase is usually not captured in full length as it is longer than the total observation time, and consequently, the measured *interphase* durations would not provide meaningful information. Even though the underlying sample numbers are relatively low, the standard deviations (displayed as colored shadows) are in an acceptable range. Comparing the curves for the treated and the control experiments it can be seen that an effect of the treatment is visible for *prophase*, as it has been expected for depletion of PNUTS, and likewise for *metaphase*. Such plots can reveal overall effects of a treatment at one glance and thus are of high importance for the biological evaluation.

4.3.4 Summary and conclusion

We presented experimental results for the analysis of multi-cell 3D image sequences using our approach for automatic image analysis. We showed that our approach, consisting of segmentation, tracking, slice selection, static and dynamic feature extraction, classification, and phase sequence processing allows classifying cell nuclei into seven cell cycle phases, and enables robust determination of the phase durations.

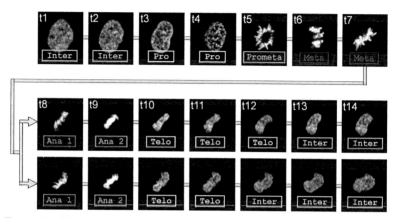

Figure 4.8: Example result for one cell nucleus: phase sequence for 14 consecutive time steps with a cell division at time step t8.

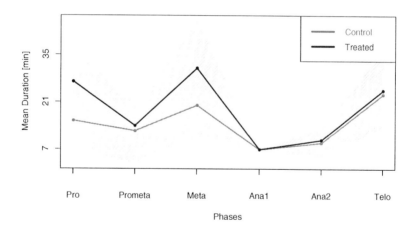

Figure 4.9: Mean phase durations for 16 control and 6 treated cell lineage trees. Standard deviations are displayed as shaded regions. The mean values are based on the following sample numbers for the control and treated experiments: prophase (14 control, 7 treated), prometaphase (14 control, 6 treated), metaphase (15 control, 7 treated), early anaphase (8 control, 3 treated), late anaphase (28 control, 10 treated), and telophase (27 control, 10 treated).

True	Classifier Output							Accur.
Class	Inter	Pro	Prom.	Meta	Ana1	Ana2	Telo	[%]
Inter	**2955**	14	0	0	0	0	17	99.0
Pro	7	**43**	1	0	0	0	0	84.3
Prom.	0	1	**34**	1	0	0	0	94.4
Meta	2	4	8	**60**	0	0	2	79.0
Ana1	0	0	2	3	**8**	1	0	57.1
Ana2	0	0	0	3	1	**43**	3	86.0
Telo	41	0	0	1	0	4	**106**	69.7

Table 4.17: Confusion matrix for classification using two-fold cross validation based on the training and test sets given in Tab. 4.16; overall accuracy: 96.6% (3249/3365).

First we demonstrated that including dynamic image features leads to a significant improvement of the classification accuracy, in particular, for the most challenging class *early anaphase*. Next, we compared different feature selection strategies regarding the classification performance on the reduced feature sets. Classification based on the selected feature subsets consisting of 50 to 150 features yielded overall accuracies slightly below the accuracy achieved based on the whole set of 355 features. Here, the methods SAM (significance analysis of microarrays) and stepwise discriminant analysis performed somewhat better than the classification accuracy-based method. For larger data sets the slightly decreased overall accuracy may be acceptable given the decrease of computation time for the reduced number of features. However, the quality of the feature selection naturally depends on the size of the data set used for feature selection.

Finally, we validated the performance of our phase sequence parser based on a subset of 22 cell lineage trees including samples from treated as well as control experiments. The evaluation of the error correction performance showed that the phase sequence parser robustly corrects inconsistencies, and thus allows fully automatic determination of the phase durations with high accuracy. All phase durations of the analyzed cell lineage trees as well as the mean phase durations were computed. As a final result we showed that the mean phase durations for prophase and metaphase were increased for the PNUTS depleted experiments compared to the control experiments.

The results described in this section have been published in conference proceedings [74, 75, 76, 77], and in a book chapter [79].

4.4 Analysis of 3D Image Sequences with Abnormal Morphologies

4.4.1 Mitosis perturbation induced by small molecule drug treatment and siRNA treatment

The images analyzed in this section are also related to the MitoCheck secondary screens. Here, two sets of proof-of-principle experiments were performed to validate our approach for automatic determination of cell cycle phase durations. Mitotic progression was perturbed using different doses of the microtubule depolymerizing drug nocodazole, i.e. 8 nmol (*low*), 10 nmol (*medium*), and 12 nmol (*high*), and alternatively by siRNA depletion of *ch-TOG* (also known as *CKAP5*), a microtubule-associated protein (MAP) involved in spindle organization in diverse organisms [42, 59]. In prometaphase, normally all chromosomes are attached, bi-oriented, and congressed into an equatorial metaphase plate by spindle microtubules. When microtubules were perturbed by either assay, chromosome congression defects occurred, resulting in a delay or even arrest in a prometaphase-like state (for simplicity, this state was labeled as prometaphase). If these defects persisted beyond prometaphase, chromosomal abnormalities could be observed: lagging chromosomes and segregation defects during anaphase, appearance of micronuclei with diverse shapes, sizes, and intensities during late anaphase as well as telophase, and multinucleated and/or abnormally shaped, non-spheroidal interphase nuclei (see Fig. 4.10).

For both assays 3D image sequences were acquired on a confocal laser scanning microscope with an image acquisition interval of seven minutes and three image slices per time step. The image slices have a spatial resolution of 1024×1024 pixels ($63\times$ objective, voxel size $0.14\mu m \times 0.14\mu m \times 3.0\mu m$) and a gray value depth of 8 bit. The total observation time varied for the different experiments between 15h and 25h. Apart from the morphological phenotypes, the images analyzed here are similar to the images considered in Sect. 4.3.

4.4.2 Approach

We applied and validated our approach for automatic determination of cell cycle phase durations described in Sect. 3.2.2 on 48 3D image sequences (36 sequences from the nocodazole experiments and 12 sequences from the ch-TOG knockdown experiments). Again, segmentation and tracking were performed based on the maximum intensity projection (MIP) images. Here we used our extended segmentation approach which enables segmentation and merging of detached chromosomes,

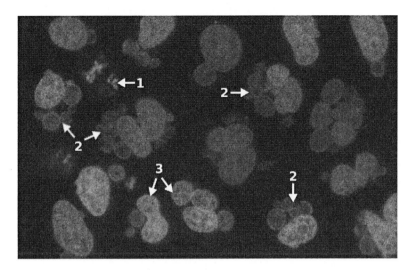

Figure 4.10: Section of a typical image from the 12 nmol (*high*) nocodazole treated experiments showing late phenotypes. The occurring phenotypes are labeled as follows: (1) detached chromosomes, (2) micronuclei, and (3) multinucleated and/or abnormally shaped nuclei.

and proper segmentation of dim micronuclei (see Sect. 3.3.5). For tracking mitotic cell nuclei, our newly developed mitosis detection criterion based on morphological properties was applied (see Sect. 3.4.3). In comparison to the previously described applications, we here computed the image features partly based on the MIP images and partly based on the most informative slices (which were selected using the maximum total intensity criterion). To determine the morphological abberations induced by the treatment (Fig. 4.10), we introduced five morphological phenotype classes in addition to the previously used seven cell cycle phases. Cell nuclei were classified into the following twelve classes: (1) *interphase*, (2) *prophase*, (3) *prometaphase*, (4) *metaphase*, (5) *early anaphase*, (6) *late anaphase*, (7) *telophase*, (8) *abnormal interphase* (including interphase nuclei with attached micronuclei, abnormally shaped and multinucleated cell nuclei), (9) *cell death*, (10) *abnormal early anaphase* (detached chromosomes), (11) *abnormal late anaphase* (detached chromosomes), and (12) *abnormal telophase* (see Fig. 4.11, for additional example images of the morphological phenotype classes (8)-(12) see Appendix Fig. A-2.1). For training of the classifier a training set including 20 image sequences (16 nocodazole and 4 ch-TOG sequences) was annotated manually by experts using the *ImageJ*-based annotation tool we developed (see Sect. 3.8.3). The numbers of sequences and samples for

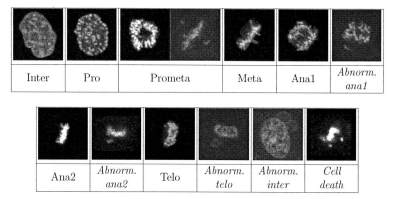

Figure 4.11: Example images for the 12 different classes, including seven normal cell cycle phases and five morphological phenotype classes (*italic*).

each class are given for the different experiments in Tab. 4.18. After classification, the resulting phase sequences were checked for consistency, corrected, and cell cycle phase durations were determined using the extended phase sequence parser (see Sect. 3.7.4).

4.4.3   Experimental results

To study the performance of the extended image analysis approach we quantified the accuracy for the segmentation as well as for the tracking algorithm. Furthermore, we analyzed the performance of different feature set compositions, using features computed based on the MIP images and on the most informative slices. Since here a relatively large feature set of 376 features was used we also performed feature reduction using principal component analysis and independent component analysis, and checked for an improvement of the overall performance. To cope with the problem of unbalanced training data we artifically downsampled the most frequent class. The results of this strategy were compared to the results obtained by weighted support vector machines which is an alternative strategy to cope with unbalanced data sets. Finally, we statistically analyzed cell cycle phase durations to determine phase prolongations or shortenings, and to detect correlations between phenotypes.

Segmentation and tracking accuracy

We evaluated the accuracy of our extended segmentation scheme based on manual inspection using four randomly picked image sequences (one for each nocodazole con-

			Inter	Pro	Prom.	Meta	Ana1	Ana2	Telo
Nocodazole	Man	Control	15649	123	111	107	43	94	556
		Low	9041	75	516	151	19	39	225
		Medium	7748	127	644	133	23	76	504
		High	9276	218	2474	211	7	18	176
	Auto	Control	40812	703	419	312	84	290	1798
		Low	4421	69	313	77	5	23	134
		Medium	2961	49	177	40	5	32	134
		High	5502	217	859	161	18	41	172
RNAi	Man	Control	6181	54	50	46	18	48	227
		Treated	2787	54	1619	52	4	9	57
	Auto	Control	9644	112	194	64	25	63	471
		Treated	6242	97	3917	25	34	32	244
Total number			120264	1898	11293	1379	285	765	4698

			Abnor. inter	Cell death	Abnor. ana1	Abnor. ana2	Abnor. telo	Number sequ.
Nocodazole	Man	Control	347	0	0	2	10	4
		Low	675	0	9	12	99	4
		Medium	1614	0	8	17	136	4
		High	4797	133	26	41	518	4
	Auto	Control	1873	4	9	14	175	14
		Low	1018	0	5	4	52	2
		Medium	757	0	0	2	32	2
		High	789	0	5	4	90	2
RNAi	Man	Control	699	0	0	1	11	2
		Treated	373	115	3	4	56	2
	Auto	Control	1665	3	0	9	20	4
		Treated	686	184	0	5	78	4
Total number			15293	439	65	115	1277	48

Table 4.18: Number of available samples per class for different data sets. "Man": manually annotated training set, "Auto": automatically annotated data set. The last column gives the number of underlying image sequences.

Experiment	Total no. of cells	Correctly segmented	Under-segmended	Over-segmented	Accur. [%]
Noco. Control	4015	3989	26	-	99.4
Noco. Low	2263	2252	10	1	99.5
Noco. Medium	3298	3296	2	-	99.9
Noco. High	5020	4774	175	71	95.1
Total No.	14596	14311	213	72	98.1

Table 4.19: Segmentation accuracy for all nocodazole treatments based on one randomly picked sequence per treatment. In particular, the proportion of undersegmetations and oversegmentations was quantified: 74.7% of the errors are undersegmentations and 25.3% are oversegmentations.

centration and one control). Segmentation accuracy was determined based on overlay images, including the transmission light channel, the GFP channel (cell nuclei), and the segmentation result. We quantified the occurrence of undersegmentations (i.e. erroneous co-segmentation of two objects as one object) and oversegmentations (i.e. erroneous segmentation of one object as multiple objects) for all nuclei at all time points, and yielded an overall accuracy of 98.1% (see Tab. 4.19). The segmentation errors of 1.9% comprised 3/4 undersegmentations and 1/4 oversegmentations.

The tracking accuracy was determined based on the same four image sequences as already used to determine the segmentation accuracy above. We found that for a total number of about 16900 matches, 40 wrong matches occurred, yielding an overall accuracy of approximately 99.8%. 29 of the 40 erroneous matches were caused by segmentation errors, 9 by mitosis detection errors, and only 2 were errors of the correspondence finding (see Tab. 4.20(a)). The mitosis detection accuracy using our new criterion was determined based on 22 image sequences with three sequences for each nocodazole concentration, six sequences for the nocodazole controls, and four sequences for the RNAi experiments (see Tab. 4.20(b)). The mitosis detection yielded an overall accuracy of 95.4% and a positive predictive value of 92.0%. Here, the 18 observed false positives were caused by abnormal morphologies such as attached micronuclei detaching from the main nucleus (which rarely happens). Even for abnormal mitotic segregation such as splits into more than two daughter chromosome sets our tracking scheme proved to work well. An example tracking result including a split into three daughters is shown in Fig. 4.12.

Classification accuracies using different feature sets

In this application we extracted features both, based on the maximum intensity projection (MIP) images, and based on the most informative slices. The reason for

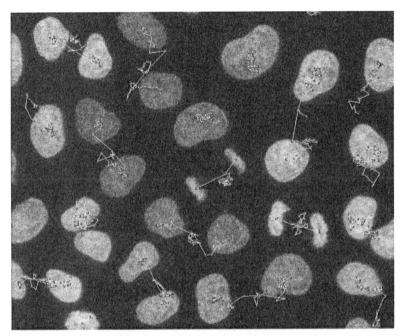

(a) Example tracking result with multiple splitting trajectories.

(b) Four consecutive time steps showing a division into three daughters.

Figure 4.12: Example results for tracking using the new morphology-based mitosis detection approach. (a) Section of a multi-cell image, (b) result for a split into three daughter chromosome sets displayed in four consecutive time steps.

a

Experiment	No. of tracks	Approx. no. of links	Errors by segm.	Errors by mitosis detect.	Errors by corresp. finding	Accur. [%]
Noco. Control	21	5000	3	-	2	99.9
Noco. Low	14	2450	1	3	-	99.8
Noco. Medium	19	3600	-	2	-	99.9
Noco. High	27	5900	25	4	-	99.5
Total No.	**81**	**16950**	**29**	**9**	**2**	**99.8**

b

Experiment	Total No.	TP No.	FP No.	PPV [%]	Sensitivity [%]
Noco Control	103	99	5	95.2	96.1
Noco Low	25	23	2	92.0	92.0
Noco Medium	32	31	1	96.9	96.9
Noco High	28	27	10	73.0	96.4
RNAi	30	28	0	100.0	93.3
Total No.	**218**	**208**	**18**	**92.0**	**95.4**

Table 4.20: Tracking accuracy. (**a**) Overall accuracy of tracking (including mitosis detection errors) based on the same four image sequences as in Tab. 4.19. The errors are subdivided into errors caused by (1) segmentation, (2) mitosis detection, and (3) correspondence finding. (**b**) Mitosis detection accuracy for all treatments based on 22 image sequences (including the four sequences used in (**a**)). Total: total numbers of occurring mitoses, TP: true positives (correctly detected), FP: false positives, PPV: positive predictive value (PPV = TP/(TP+FP)) and the sensitivity (sensitivity = TP/Total).

		Feature type	Number
Computation on MIP images		Features based on size and shape	12
		Geometric moments-based	5
		Mean intensity	1
		Tree structured wavelets-based [28]	4
		Zernike moments [195]	49
		Dynamic features (mean intensity, size, shape)	10
Computation on most inform. slices		Granularity (local differentiation)	20
		Gray scale invariants [24]	8
		Haralick texture features [72]	260
		Tree structured wavelet-based [28]	4
		Standard deviation	1
		Dynamic features (standard deviation)	2
		Total Number	**376**

Table 4.21: Feature types and feature numbers extracted for each type. 81 features were computed based on the maximum intensity projections, and 295 features were computed based on the most informative slices.

using the original image slice for feature extraction is that fine textures, which are important for the classification of certain phases (e.g., prophase), can be blurred in the MIP images. On the other hand, the MIP images contain the entire object, e.g., detached chromosomes often were not located in the most informative slice. Consequently, we performed classification experiments based on three different feature sets: (a) computed on the MIP images, (b) computed on the most informative slices, and (c) a combination of features from set (a) and (b). For the combined feature set (c) we computed 81 features that were primarily based on object shape, e.g., size, shape (circularity, Feret's diameter), contour length, or Zernike moments on the (non-smoothed) MIP images, and 295 features related to texture, like Haralick texture features, granularity features, gray scale invariants, or wavelet features on the most informative slices (see Tab. 4.21). The most informative slice was determined automatically for each cell based on maximum total intensity. The classification results for the nocodazole training data given in Tab. 4.22 are relatively similar for the different feature sets, however, the set (c) most often provided the highest single class accuracy. Furthermore, the highest overall classification accuracy of 93.9% was reached for set (c) in comparison to 93.8% and 93.7% for set (a) and (b), respectively. Consequently, we used the combined feature set (c) for classification in this application.

We trained two SVMs (with Gaussian radial basis function kernel), one for the nocodazole data and one for the ch-TOG RNAi knockdown data. The performance for each classifier was evaluated using five-fold outer cross validation based

Set	1	2	3	4	5	6	7	8	9	10	11	12	Ova	Avg
a	98.0	61.7	95.7	82.4	71.7	90.8	83.5	93.3	91.7	30.2	37.5	64.4	93.8	75.1
b	98.0	65.2	95.5	82.4	68.5	92.1	82.6	93.3	91.7	30.2	41.7	62.8	93.7	75.3
c	98.0	67.8	95.4	82.6	70.7	90.8	84.7	93.5	91.7	25.6	41.7	63.4	93.9	75.5

Table 4.22: Classification accuracies for all classes resulting from a five-fold cross validation on the nocodazole training data using different feature sets: (**a**) features from MIP image, (**b**) features from the most informative slice, (**c**) combination of both feature sets. The column numbers 1-12 represent the classes (1) inter-, (2) pro-, (3) prometa-, (4) meta-, (5) early ana-, (6) late ana-, (7) telo-, and (8) abnormal interphase, (9) cell death, (10) abnormal early ana-, (11) abnormal late ana-, and (12) abnormal telophase. The column "Ova" provides the overall accuracies and the column "Avg" the average accuracies. The interphase samples were reduced to 1000 samples per sequence. The gray tones represent the accuracy ranking within the three groups, where dark gray indicates the lowest accuracy and white the highest accuracy.

on the manually annotated training data (16 nocodazole sequences and four RNAi sequences, see Tab. 4.18). To obtain a more balanced data set and reduce the computation time for training of the classifier, the number of interphase cells was reduced to 1000 samples per sequence (see Sect. 4.4.3). We yielded overall classification accuracies of 93.9% for the nocodazole data and 94.7% for the RNAi data. The detailed confusion matrices for both classification experiments are given in Tab. 4.23. For both experiments it can be seen that most classes reached accuracies in the range of 82.6% to 98.0%. Lower accuracies in the range of 63.0% to 70.7% were yielded for the classes prophase, early anaphase, and abnormal telophase. Only for the classes abnormal early and late anaphase the resulting accuracies were in the range of 0.0% to 41.7%. However, note that in this group the sample numbers were extremely low, in particular, for the RNAi data set (3 and 5 samples). In addition, especially the low accuracy classes are characterized by a very high intra-class variability and a high similarity to related classes (e.g., abnormal class and normal class, or preceding and succeeding phase, see Appendix Fig. A-2.1). The final classifiers trained on the complete training data were applied to previously unseen test data consisting of 20 sequences for nocodazole and eight sequences for RNAi experiments, respectively.

Classification after feature reduction

To study the effect of feature reduction we performed classification experiments on reduced feature sets. To this end, principal component analysis (PCA) and independent component analysis (ICA) were applied to the original feature set (see

a

True Class	Classifier Output											
	1	**2**	**3**	**4**	**5**	**6**	**7**	**8**	**9**	**10**	**11**	**12**
1	**15674**	40	1	0	1	0	113	169	0	0	0	2
2	151	**368**	15	0	0	0	1	8	0	0	0	0
3	7	14	**3572**	93	4	3	3	12	2	9	5	21
4	2	0	85	**497**	6	0	4	1	2	0	1	4
5	0	0	10	7	**65**	2	1	0	0	7	0	0
6	0	0	1	2	4	**206**	3	0	1	0	10	0
7	139	0	1	6	2	2	**1238**	13	0	0	2	58
8	369	5	7	1	0	0	32	**6951**	0	0	0	68
9	3	1	5	0	0	0	0	0	**122**	0	2	0
10	0	0	18	0	11	0	0	0	0	**11**	2	1
11	0	0	15	3	0	15	3	0	2	1	**30**	3
12	12	0	32	3	0	2	111	117	1	0	1	**484**
Acc. [%]	98.0	67.8	95.4	82.6	70.7	90.8	84.7	93.5	91.7	25.6	41.7	63.4

b

True Class	Classifier Output											
	1	**2**	**3**	**4**	**5**	**6**	**7**	**8**	**9**	**10**	**11**	**12**
1	**3825**	3	0	0	0	0	16	87	0	0	0	0
2	30	**68**	6	0	0	0	0	4	0	0	0	0
3	0	3	**1646**	10	0	2	1	0	3	0	0	4
4	0	0	9	**85**	1	0	3	0	0	0	0	0
5	0	0	4	1	**15**	2	0	0	0	0	0	0
6	0	0	1	0	0	**54**	2	0	0	0	0	0
7	21	0	2	0	0	3	**244**	10	0	0	0	4
8	99	2	0	1	0	0	6	**959**	1	0	0	4
9	1	0	10	1	0	1	0	1	**100**	0	0	1
10	0	0	3	0	0	0	0	0	0	**0**	0	0
11	0	0	3	1	0	1	0	0	0	0	**0**	0
12	2	0	8	0	0	0	7	6	0	0	0	**44**
Acc. [%]	97.3	63.0	98.6	86.7	68.2	94.7	86.0	89.5	87.0	0.0	0.0	65.7

Table 4.23: Confusion matrix for SVM classification using five-fold cross validation on the training set (the interphase samples were reduced to 1000 samples per sequence), (**a**) for the nocodazole data, overall accuracy: 93.9%, and (**b**) for the RNAi data, overall accuracy: 94.7%. The numbers 1-12 denote the classes (1) interphase, (2) prophase, (3) prometaphase, (4) metaphase, (5) early anaphase, (6) late anaphase, (7) telophase, (8) abnormal interphase, (9) cell death, (10) abnormal early anaphase, (11) abnormal late anaphase, and (12) abnormal telophase. The last row provides the classification accuracies per class in percent.

Sect. 3.5.4). First, we applied PCA using different numbers of principal components and compared the classification results with the result using all original features. Note that here a combined classifier was trained based on data from both, the nocodazole and the RNAi experiments, which performed similar to the separate classifiers (compare Tab. 4.24(a)) and Tab. 4.23). It turned out that for the reduced feature sets of 5, 10, 20, 30, 50, and 100 principal components we obtained significantly lower classification accuracies than for the original feature set (see Fig. 4.13). For 50 and 100 principal components we obtained the highest classification accuracies of 92.6% and 92.3%, respectively, in comparison to 94.0% for the original feature set. Table 4.24 shows the confusion matrices for classification based on the original feature set and based on 100 principal components. Additionally, we performed experiments using independent component analysis (ICA). 50 and 100 independent components resulted in very similar results as for PCA (92.4% and 92.3%, respectively; see Appendix Tab. A-1.1).

In conclusion, both feature reduction techniques neither increased the overall classification accuracies, nor increased any of the single class accuracies. Particularly for the low accuracy classes (e.g., abnormal early and late anaphase) the performance was decreased significantly (see Tab. 4.24). On the other hand, however, the computation time for training of the classifier was reduced for the reduced feature sets. Since in this application a high classification accuracy was more desirable than the moderate gain in the total processing speed, we used the original feature set in the following processing steps.

Classification using weighted support vector machines

As can be seen in Tab. 4.18 our training data set is highly unbalanced, in particular, the sample number for the most frequent class (interphase) is about 2000 times higher than the sample number for the least frequent class (abnormal early anaphase). Such unbalanced training data can decrease the classification performance for the less frequent classes. Common strategies to deal with this problem are resampling the training data set (to reduce the sample number for the frequent classes or to increase the number for the less frequent classes) or using weighted classifiers. In our case we decreased the interphase sample number by downsampling to 1000 samples per sequence. Here, downsampling of the interphase samples is an adequate strategy since there are many redundant samples within this class. If a cell stays in interphase and does not migrate, the nucleus remains unchanged for many time steps, and thus, we have a large number of almost identical interphase training nuclei which can be reduced without significant loss of information. We

a												
True	Classifier Output											
Class	1	2	3	4	5	6	7	8	9	10	11	12
1	19494	45	1	1	1	0	143	242	0	0	0	4
2	163	452	22	0	0	0	3	11	0	0	0	0
3	8	21	5214	110	3	3	4	9	6	12	6	18
4	1	1	88	586	11	0	6	0	2	0	1	4
5	0	0	11	10	83	2	1	0	0	7	0	0
6	0	0	2	1	4	260	7	0	1	0	9	0
7	171	0	4	6	1	5	1475	21	0	0	3	59
8	478	6	8	2	1	0	43	7891	1	0	0	75
9	4	0	15	3	0	1	0	1	221	0	2	1
10	0	0	20	0	9	0	0	0	0	14	3	0
11	0	0	19	3	1	13	3	0	2	1	32	3
12	10	0	35	4	0	2	125	124	0	0	2	528
Acc. [%]	97.8	69.4	96.3	83.7	72.8	91.6	84.5	92.8	89.1	30.4	41.5	63.6

b												
True	Classifier Output											
Class	1	2	3	4	5	6	7	8	9	10	11	12
1	19411	33	2	0	0	0	153	327	0	0	0	5
2	200	420	20	1	0	0	3	7	0	0	0	0
3	8	20	5242	95	6	3	2	11	2	1	5	19
4	8	1	92	577	2	2	7	2	0	1	4	4
5	0	0	14	12	76	6	0	0	0	4	2	0
6	2	0	3	0	3	258	11	0	0	0	6	1
7	205	0	2	4	1	4	1454	23	0	0	1	51
8	917	5	19	1	2	0	52	7435	1	0	0	73
9	5	0	22	0	0	0	0	1	218	0	1	1
10	0	0	23	1	9	0	1	0	0	10	0	2
11	2	0	23	3	2	16	6	0	0	1	22	2
12	28	0	37	4	0	1	161	152	0	0	32	444
Acc. [%]	97.4	64.5	96.8	82.4	66.7	90.9	83.3	87.4	87.9	21.7	28.6	53.5

Table 4.24: Confusion matrix for the combined classifier trained on the nocodazole and the RNAi data using five-fold cross validation (the interphase samples were reduced to 1000 samples per sequence). (a) Classification based on the complete feature set, overall accuracy: 94.0%, (b) classification based on 100 principal components, overall accuracy: 92.3%.

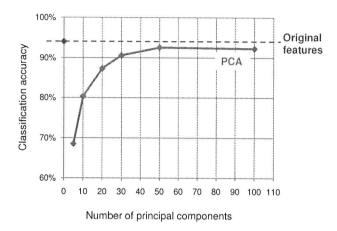

Figure 4.13: Classification accuracies for 5, 10, 20, 30, 50, and 100 principal components (solid line), and classification accuracy for the original feature set (dashed line). Here, combined classifiers were used, i.e. trained on the nocodazole and the RNAi data.

selected the reduced set of interphase samples using systematic sampling, i.e. each nth interphase sample is picked from the sequence of interphase samples (n is chosen for each image sequence as the total number of interphase samples divided by 1000, see, e.g., [84]).

We also investigated weighted support vector machines [139] (see Sect. 3.6.2) as an alternative technique to deal with the unbalanced training data. To check whether weighted SVMs yield higher classification accuracies than normal SVMs using the resampled training data (as described above), we repeated the five-fold cross validation on the RNAi data set, using all available interphase samples and weighted SVMs. Like in all other experiments we used the SVM library LIBSVM [27], which also provides an implementation of weighted SVMs. The weights were determined for class i as the number of samples for the largest class (here interphase) divided by the number of samples for class i, as proposed, e.g., in [179]. The result is shown in Tab. 4.25 (compare to Tab. 4.23(b)). It can be seen that for most classes with small sample size the accuracy decreased (partially significantly, e.g., for early anaphase by 9.1% and abnormal interphase by 8.9%). Only for prometaphase and metaphase the accuracy increased, but only by 0.1%. The accuracy for interphase increased by 1.0% which resulted in an overall accuracy of 95.6%, compared to the overall accuracy of 94.7% as yielded for the resampled training data. Consequently,

True Class	Classifier Output											
	1	2	3	4	5	6	7	8	9	10	11	12
1	**8811**	5	0	0	1	0	29	122	0	0	0	0
2	35	**64**	6	0	0	0	0	3	0	0	0	0
3	2	3	**1648**	8	0	2	1	2	2	0	0	1
4	0	0	8	**86**	1	0	3	0	0	0	0	0
5	1	0	5	1	**13**	2	0	0	0	0	0	0
6	0	0	1	1	0	**52**	3	0	0	0	0	0
7	32	0	2	0	0	3	**234**	8	0	0	0	5
8	196	2	1	1	0	0	3	**864**	1	0	0	4
9	2	0	10	1	0	0	0	1	**100**	0	0	1
10	0	0	3	0	0	0	0	0	0	**0**	0	0
11	0	0	4	0	0	1	0	0	0	0	**0**	0
12	6	0	7	0	0	0	7	5	0	0	0	**42**
Acc. [%]	98.3	59.2	98.7	87.8	59.1	91.2	82.4	80.6	87.0	0.0	0.0	62.7

Table 4.25: Confusion matrix for a weighted SVM classifier trained on the RNAi data using five-fold cross validation, overall accuracy: 95.6%.

the weighted SVM scheme did not significantly improve the accuracies for classes with small sample numbers in our case. The improvement of the overall accuracy was almost exclusively caused by the increased interphase accuracy.

Statistical analysis of cell cycle phase durations

The phase durations for all experiments were determined using our extended phase sequence parser, resulting in phase duration distributions (histograms) for the different experiments (see Fig. 4.26, see also Appendix Fig. A-2.2). To prove that our approach allows accurate determination of changes in mitotic progression, we analyzed the effect of nocodazole and siRNA knockdown treatment on the durations of mitotic phases with particular attention to prometaphase. We tested whether prometaphase in perturbed cells was significantly longer than in controls, and compared the dose response of three different nocodazole concentrations (low, medium, and high). Furthermore, we tested whether the results for the automatically annotated data and for the manually annotated data were significantly different.

Statistical analysis For a sound statistical analysis of phase duration differences we compared the distributions of phase durations over data sets using statistical testing instead of simply comparing mean values. To determine an appropriate statistical test we first checked whether the phase durations were normally distributed,

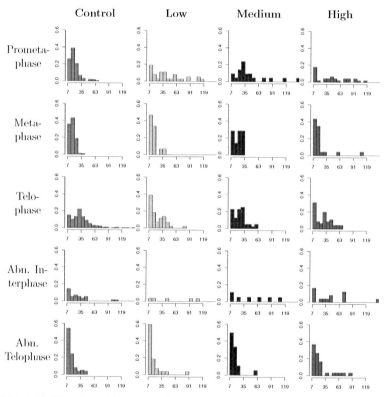

Table 4.26: Phase length histograms for automatically annotated data. x-axis: phase length [min], y-axis: relative frequency. Histograms are displayed for a maximum length of 140 minutes (20 time steps).

using the Shapiro-Wilk test [164] for all experiments. Since for almost all experiments a normal distribution could not be assumed (see Appendix Tab. A-1.2), we selected a non-parametric test, in particular, the Mann-Whitney U test [122] for not paired data. One-sided Mann-Whitney tests were applied in both directions to test whether significant shifts between phase duration distributions existed (e.g., between treated and control experiments, or between different treatment concentrations; alternative hypotheses: "true shift is greater/smaller than 0"). To check whether significant differences existed between the manually and the automatically annotated data we applied two-sided Mann-Whitney tests (alternative hypothesis: "true shift is not equal to 0"). For all statistical tests we used a significance level of $\alpha = 5\%$. Statistical analysis was performed using R, the language and environment for statistical computing [154].

Phase prolongations and shortenings The comparison of the prometaphase length distributions of nocodazole treated and control experiments revealed a highly significant prometaphase prolongation for all nocodazole concentrations (see Tab. 4.27, see also Appendix Fig. A-2.3). 12 nmol (high) nocodazole concentration showed a stronger prolongation w.r.t. the control experiment (difference of mean prometaphase durations ca. 127 min) of higher significance ($p_{high} = 4.2 \cdot 10^{-12}$, Mann-Whitney test), compared to the effect for concentrations of 10 nmol and 8 nmol (medium and low, difference of means for both ca. 45 min, $p_{low} = 1.1 \cdot 10^{-7}$ and $p_{medium} = 4.1 \cdot 10^{-7}$), respectively. The medium concentration produced more cases of strong delay, but the low concentration datasets gave a higher proportion of delayed cells, showing a dose response. Regarding prometaphase duration, however, medium and low concentration of nocodazole were not significantly different. The automatic analysis and statistical evaluation furthermore showed that cells depleted for ch-TOG siRNA had an even stronger increase of prometaphase duration (difference of means ca. 218 min, $p_{RNAi} = 2.3 \cdot 10^{-12}$, Mann-Whitney test) compared to control cells treated with scrambled (scr) siRNA (see Tab. 4.27, see also Appendix Fig. A-2.3). In Fig. 4.14 this analysis is summarized in a compact manner by plotting the mean phase durations for all normal and abnormal phases. This plot reveals at one glance that the perturbations are specific to prometaphase and that the increasing strengths of the phenotypes are caused by rising doses of nocodazole and ch-TOG siRNA.

To detect effects of the treatments on other phase durations, we first performed two-sided Mann-Whitney U tests for all phases to determine whether a shift between the treated and control distributions existed. If so, one-sided tests were performed to determine the direction of the shift (i.e., phase shortening or prolongation). For the

	Nocodazole						RNAi
Man	Low-Control	Medium-Control	High-Control	Medium-Low	High-Medium	High-Low	Treated-Control
n_1, n_2	44, 57	68, 57	120, 57	68, 44	120, 68	120, 44	29, 27
p-value	$4.6 \cdot 10^{-12}$	$2.6 \cdot 10^{-13}$	$2.2 \cdot 10^{-16}$	0.98	$3.3 \cdot 10^{-7}$	$2.1 \cdot 10^{-2}$	$8.0 \cdot 10^{-9}$
Auto	Low-Control	Medium-Control	High-Control	Medium-Low	High-Medium	High-Low	Treated-Control
n_1, n_2	37, 180	21, 180	45, 180	21, 37	45, 21	45, 37	115, 66
p-value	$1.1 \cdot 10^{-7}$	$4.1 \cdot 10^{-7}$	$4.2 \cdot 10^{-12}$	0.70	$9.1 \cdot 10^{-3}$	$1.1 \cdot 10^{-2}$	$2.3 \cdot 10^{-12}$

Table 4.27: Results of Mann-Whitney U tests on prometaphase length distributions for all experiments. One-sided tests with the alternative hypothesis: "true shift is greater than 0" were used, n_1 and n_2 are the numbers of the analyzed prometaphase sequences.

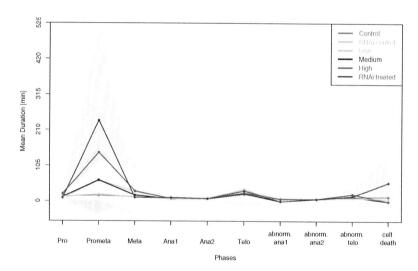

Figure 4.14: Mean phase durations for the automatically annotated data for all treatments and controls. The shaded regions indicate the standard deviations.

nocodazole experiments we found a significant shortening of interphase for all concentrations in the manually annotated data and for low and medium concentrations
in the automatically annotated data, while abnormal interphase was shortened significantly only in the manually annotated data. However, this does not necessarily
indicate true differences in interphase length since cells under nocodazole treatment
were strongly delayed in prometaphase while the total duration of the time-lapse
experiments was the same as for the controls. Consequently, many interphases were
not recorded until the end and may appear shorter. Late anaphase was significantly
prolonged for the manually annotated data. For telophase we detected a significant
shortening for all concentrations in the manually and automatically annotated data.
This effect was caused by alternations between telophase and abnormal telophase.
Finally, a significant prolongation was detected for abnormal telophase for low and
high concentrations in the manually annotated data and for high concentrations
in the automatically annotated data (see Appendix Tab. A-1.3(a)). For the RNAi
treated experiments we found a significant shortening of abnormal interphase and
telophase for the automatically annotated data which occurred due to the same
reasons as described above. Also, we detected a significant prolongation of early
anaphase and late anaphase in the automatically annotated data and a significant
prolongation of abnormal telophase in the manually annotated data (see Appendix
Tab. A-1.3(b)). Note that here the sample numbers were comparatively low. For
both types of experiments the statistical tests could not be performed for apoptosis
and abnormal early anaphase due to small sample numbers.

Comparison of manually and automatically annotated data To check the
reliability of the automatic annotation we investigated whether the same overall
results were yielded for the manual and automatic annotation. The cumulative
histograms in Fig. 4.15 already indicate that the results are similar. In addition,
Mann-Whitney U tests were performed for all experiments between the phase length
distributions of the manually and automatically annotated data. For most experiments we found no significant differences, however, for the nocodazole control and
an the RNAi treated experiments the resulting p-values were slightly below the
significance threshold (see Appendix Tab. A-1.4). To further clarify this issue we
re-examined the corresponding source data. For the nocodazole control experiments
we found one outlier image sequence with prolonged prometaphase (low L10, automatically annotated, see Appendix Fig. A-2.4) which usually does not occur in
non-perturbed cells. Manual verification showed that the observed effect was present
in the data, which consequently most likely was caused by an experimental error. Af-

ter discarding the respective sequence the Mann-Whitney test resulted in a p-value of 0.07, indicating no significant difference between manually and automatically annotated data. For the treated RNAi experiments no outlier could be identified (see Appendix Fig. A-2.4). However, the variability of the data is relatively high given the small number of sequences. Also, non-corrected classification errors within long prometaphase sequences can cause splits into two shorter prometaphase sequences. Thus, few long prometaphase sequences were erroneously counted as two shorter prometaphase sequences in the automatically annotated data, resulting in a slightly less prominent prometaphase prolongation.

Correlation analysis between phenotypes

To test whether prometaphase delays were linked with chromosome segregation and nuclear shape abnormalities (including cell death) in later stages of mitosis, we performed a temporal correlation analysis. First, we determined the percentage of prolonged prometaphases in the different spindle perturbation experiments (see Appendix Fig. A-2.5). As a threshold for normal prometaphase length the mean plus twice the standard deviation of prometaphase duration in the control cells was used (mean ca. 15 min +/- 11 min; see Appendix Tab. A-1.5). To calculate the probability of the occurrence of morphological phenotypes after prometaphase, we divided the number of time steps with abnormal morphologies occurring after prometaphase by the total number of time steps after prometaphase for each mitotic cell, which was denoted as ratio r. This analysis showed that prolonged prometaphases, caused by chromosome congression and alignment defects, resulted in a higher number of aberrant late mitotic phenotypes, which was reflected by a higher probability compared to normal prometaphases (see Appendix Tab. A-1.6).

Next, we determined correlation coefficients between prometaphase duration and late phenotype probability r. We used the non-parametric correlation coefficients Kendall's τ and Spearman's ρ since the data could not be assumed to be normally distributed (as discussed above). Scatter-plots of the data indicated that the correlation coefficient was influenced by cases with $r = 0$ (see Appendix Fig. A-2.6). Since these cases occur, e.g., when prometaphase is cut off at the end of an image sequence, they were excluded for the correlation analysis. We obtained values of $\tau = 0.38$ and $\rho = 0.51$ for the manually annotated data, and $\tau = 0.27$ and $\rho = 0.38$ for the automatically annotated nocodazole data. For the RNAi experiments we yielded generally higher values of $\tau = 0.63$ and $\rho = 0.73$ for the manually annotated data, and $\tau = 0.43$ and $\rho = 0.57$ for the automatically annotated data. All results proved to be highly significant (see Tab. 4.28). Note that even when including

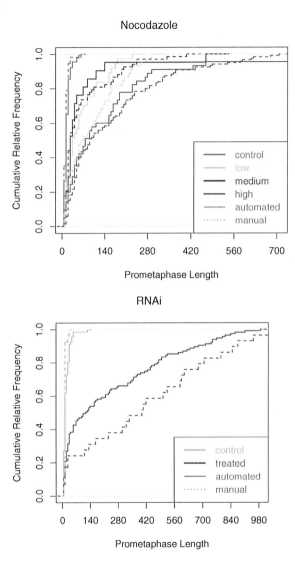

Figure 4.15: Cumulative histograms of prometaphase duration in minutes for all concentrations of the nocodazole experiments (top), and for the scrambled control and ch-TOG RNAi experiments (bottom). Solid lines represent automatically annotated data and dashed lines manually annotated data.

		Nocodazole		RNAi	
		Manual	**Auto**	**Manual**	**Auto**
n		110	136	15	69
Kendall	τ	0.38	0.27	0.63	0.43
	p-value	$9.5 \cdot 10^{-9}$	$8.6 \cdot 10^{-6}$	$1.3 \cdot 10^{-3}$	$4.2 \cdot 10^{-7}$
Spearman	ρ	0.51	0.38	0.73	0.57
	p-value	$7.8 \cdot 10^{-9}$	$2.5 \cdot 10^{-6}$	$9.4 \cdot 10^{-4}$	$2.1 \cdot 10^{-7}$

Table 4.28: Correlation coefficients τ and ρ for prometaphase duration and ratio r (probability of following abnormal morphologies), excluding cases with ratio $r = 0$. Additionally, the p-values for the significance tests and numbers of measurements n are given.

the exceptional cases with $r = 0$, in most cases the correlations were still significant (see Appendix Tab. A-1.7). For the significance tests we used significance level $\alpha = 5\%$. This temporal correlation of different phenotypes shows quantitatively and with statistical significance that even mild perturbations of the spindle that result in transient prometaphase delays cause chromosome segregation defects, and thus allows us to put different events in a causal order in time.

Application to images of a different cell line and from a different screening platform

To demonstrate the applicability of our approach for different experimental settings we further analyzed images of a different cell line and images acquired with a different microscopy screening platform. In a first study, we used a different cell line, namely *normal rat kidney* (NRK) cells, which were imaged using the same screening platform (LSM 510 point-scanning microscope) as for the previously analyzed HeLa cells (see Fig. 4.16). We applied our image analysis approach to six image sequences of non-treated NRK cells including about 4000 chromosome sets. Our approach could be directly applied without changing any parameter value except for the maximum displacement. Since NRK cells have a higher motility than HeLa cells, for tracking we increased the parameter value for the maximum displacement. Our evaluation showed that we yielded high accuracies for segmentation (99.3%) and tracking (99.7% correspondence finding, 87.5% mitosis detection) (see Appendix Tabs. A-1.8 and A-1.9). For classification we obtained an overall accuracy of 92.9% using five-fold cross validation on all six image sequences (ground truth was generated by manual annotation; see Tab. 4.29, and Appendix Tab. A-1.10). Finally, we determined the cell cycle phase durations based on the automatically classified

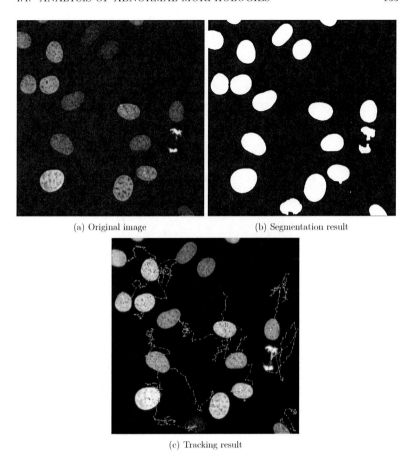

(a) Original image   (b) Segmentation result

(c) Tracking result

Figure 4.16: Example images for NRK (normal rat kidney) cells acquired on an LSM 510 Meta point scanning microscope. (a) Original image, (b) segmentation result, and (c) tracking result (red dots: centers of gravity, yellow lines: trajectories).

chromosome sets (see Fig. 4.17(a)).

In a second study, we used a different microscopy screening platform, namely the LSM 5 LIVE line-scanning microscope with a charged coupled device, which allows faster image acquisition but provides half of the resolution in the x-y direction compared to the LSM 510 Meta point-scanning microscope with a photomultiplier. Image sequences of non-treated HeLa cells were acquired on the line-scanning microscope, and we applied our approach to four image sequences including about 9100 chromosome sets. For accurate analysis of these images we only changed few

Cell line	Microscope	Number of cells	Segmen-tation	Tracking (Mitosis detection)	Classfi-cation
HeLa	point-scan. LSM 510 Meta	14596 (31114)	98.1%	99.8% (95.4%)	93.9%
HeLa	line-scan. LSM 5 LIVE	9226	99.8%	99.9% (100.0%)	97.8%
NRK	point-scan. LSM 510 Meta	4008	99.3%	99.7% (87.5%)	92.9%

Table 4.29: Overview of the accuracies of our segmentation, tracking, and classifi-cation approaches for different experimental settings. Note that for the experiment with HeLa cells imaged on a point-scanning microscope (first row) the classification accuracy was determined on a larger data set of 31114 samples compared to the segmentation and the tracking accuracies which were determined on 14596 samples.

parameter values of our approach (maximum displacement, window size, minimum nucleus size and maximum fragment distance) which were straightforwardly ob-tained by linearly scaling the values according to the change in resolution. From the evaluation we found that we yielded accuracies of 99.8% for segmentation and 99.9% for tracking (using ground truth from manual evaluation, see Appendix Tabs. A-1.8, A-1.9), and 97.8% for classification (using five-fold cross validation based on man-ually annotated ground truth, see Tab. 4.29 and Appendix Tab. A-1.10). Again, we successfully determined the cell cycle phase durations automatically for all se-quences (see Fig. 4.17(b)). An overview of the results for the different experimental settings is given in Tab. 4.29.

Experimental comparison with other segmentation and tracking approaches

Finally, we also performed a quantitative experimental comparison of our segmen-tation and tracking approaches with other approaches (based on 4 image sequences of HeLa cells, i.e. control, low, medium, and high concentration, including 14596 cell nuclei). For segmentation we applied two other often used approaches for cell nucleus segmentation, namely global Otsu thresholding and K-Means clustering (us-ing 3 clusters, which was the optimal setting for the considered data), which are included in the public domain software ImageJ [156]. We used the same prepro-cessing (i.e. Gaussian filtering) and postprocessing (i.e. hole filling) steps as for our approach. From the evaluation study we found that global Otsu thresholding yielded an accuracy of 68.9% and K-Means clustering an accuracy of 73.9%, while our ap-

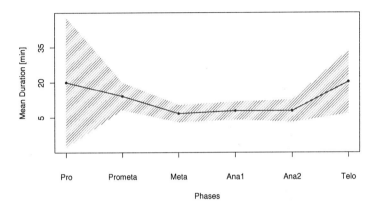

(a) NRK cells acquired on LSM 510 Meta

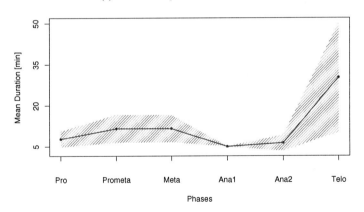

(b) HeLa cells acquired on LSM 5 LIVE

Figure 4.17: Automatically determined mean phase durations for (a) NRK (normal rat kidney) cells acquired with an LSM 510 Meta point-scanning microscope (six image sequences), and (b) HeLa cells acquired with an LSM 5 LIVE line-scanning microscope (four image sequences).

a					
Method	Correct	Under-segmented	Over-segmented	Partly/not segmented	Accuracy [%]
Global Otsu	10055	12	1516	3007	68.9
K-Means	10787	1184	1298	1327	73.9
Our method	14311	213	72	0	98.1

b					
Method	No. of links	Errors segm.	Errors mitosis detect.	Errors corresp. finding	Accuracy [%]
Cell Profiler	16956	31	43	3	99.6
MTrack2	16942	29	43	2	99.6
Our method	16950	29	9	2	99.8

Table 4.30: Comparison of our segmentation and tracking approaches with other approaches based on four image sequences with different treatments (low, medium, and high nocodazole concentration, and control sequence). The total number of analyzed cell nuclei was 14596. (a) Segmentation accuracies of our approach in comparison with global Otsu and K-Means clustering (3 clusters). (b) Tracking accuracies of our approach compared to tracking approaches in CellProfiler and ImageJ (Plugin MTrack2).

proach resulted in 98.1%. One main reason why our approach yields a significantly better result is that we cope with the issue of merging detached chromosomes. In contrast, global Otsu thresholding and K-Means clustering lead to a relatively high number of oversegmentations, as can be seen from Tab. 4.30(a). For tracking we used the cell tracking algorithm provided in CellProfiler [25] as well as the often used ImageJ plugin *MTrack2* [166]. As input for tracking we used the segmentation results from our segmentation approach. Note that both tracking algorithms cannot handle splitting events. From the analysis it turned out that for CellProfiler as well as for MTrack2 the percentage of correct correspondences was slightly lower than the result for our approach (99.6% vs. 99.8%, see Tab. 4.30(b)). More importantly, the two approaches cannot handle splitting events while a main advantage of our approach is the detection of mitosis events and the tracking of cell divisions.

4.4.4   Summary and conclusion

In this section, we presented results for analyzing multi-cell 3D image sequences with morphological phenotypes applying our extended approach. For two proof-of-principle experiments we showed that our approach enables robust, automatic determination of cell cycle phase prolongations or shortenings. First, we systematically validated the extended segmentation and tracking approach by comparing the results with manually assigned ground truth. Our segmentation as well as our tracking approach turned out to be very accurate. For segmentation we yielded accuracies of 98.1%, and for tracking we obtained 99.8% (one-to-one correspondence finding step) and 95.4% (mitosis detection step). We also compared the performance of our segmentation and tracking approaches with other commonly used segmentation and tracking approaches. For segmentation as well as for tracking our approach yielded the best results. Next, different strategies for feature extraction from the 3D images were investigated and compared. We showed that a combined feature set including features based on both, the most informative slice *and* the maximum intensity projection is superior to feature sets including features based on only one of both image type. Furthermore, the effect of feature reduction on the classification accuracy was studied applying principal component analysis and independent component analysis. For training sets including different numbers of principal and independent components (in the range of 5 to 100), respectively, the classification accuracies were significantly decreased to 68.5% to 92.6%, compared to 94.0% for the original feature set. Consequently, to yield accurate results we used the original set of 376 features. To deal with the unbalanced training data set we applied and compared two strategies: (1) artificially downsampling the number of interphase samples, and (2) applying weighted support vector machines. Here, the downsampling strategy yielded better results for most of the classes, and additionally reduced the computation time.

Using our approach we determined the cell cycle phase durations for a data set of 48 image sequences including 134 to 219 time steps each. Statistical tests were applied on the phase duration distributions to detect prolongations or shortenings in the treated experiments compared to the controls. Based on the automatically analyzed data we detected prometaphase prolongations of different extent for the different treatments, which had been expected given the experimental treatment. This proves the applicability and effectivity of our approach. Furthermore, we detected less prominent prolongations and shortenings for other cell cycle phases which were partly artefacts and partly additional treatment effects. A statistical comparison of the results for the manually and automatically annotated data showed that there were only very few significant differences between the results. This demonstrates that our automatic approach in almost all cases yielded the same overall analysis result as manual annotation. Finally, the temporal correlation between prometaphase prolongation and morphological phenotypes was quantitatively analyzed. We found a medium positive correlation between both phenotypes, which again was the expected result, proving the effectiveness of our method. We also showed that our approach can be readily applied to images acquired under different experimental settings. Using images of a different cell line and images acquired on a different microscopy platform we found that our approach yielded comparably high accuracies as for the initial application. The results described in this section have been published as a journal paper [80].

4.5 Analysis of High-Throughput 3D Image Sequences

4.5.1 RNAi screen on nuclear envelope proteins

In this section we present results of the automatic analysis of a complete siRNA knockdown screen using 104 nuclear envelope proteins. The nuclear envelope is an important membrane structure within the cell, separating DNA replication and transcription processes within the nucleus from mRNA translation in the cytoplasm. Nuclear envelope proteins mediate the exchange of molecules between nucleus and cytoplasm through nuclear pores. Moreover, several nuclear envelope proteins have additional functions during mitosis, e.g., they are involved in mitotic spindle assembly. The focus of the high-content screen analyzed in this work was to study the influence of nuclear envelope proteins on chromosome segregation and nuclear structure (for more details on the screen see [188]).

For a detailed analysis of the cell nucleus structure and mitotic phenotypes over time, image sequences with high spatio-temporal resolution were acquired. Using a confocal fluorescence microscope, five image slices per time step were taken at each spot with a spatial resolution of 512×512 pixels ($40 \times$ objective, voxel size $0.33 \mu m \times 0.33 \mu m \times 2.5 \mu m$) and a gray value depth of 8 bit. The image acquisition interval was five minutes for a total observation time of 18 hours (imaging started between 20 and 100 hours after transfection). For each of the 104 genes three replicate experiments were performed and different siRNA mixes were applied, which resulted in a total number of 951 experiments (including approximately 20% control experiments). Acquiring one image sequence with the above described resolution for each experiment resulted in a total data volume of about 250 GByte (951 sequences, 216 images per sequence, 5 slices per image, 512×512 pixels per slice with 1 Byte per pixel).

4.5.2   Approach

To analyze these images we applied basically the same workflow as used for analyzing 3D image sequences with abnormal morphologies as described in Sect. 4.4 above. As mentioned there, the extended segmentation approach was used to enable accurate segmentation of morphological phenotypes (see also Sect. 3.3.5). For tracking, again the enhanced mitosis detection approach was used (see Sect. 3.4.3). However, here the image features were computed solely based on the maximum intensity projection (MIP) images because of the very large number of images. For each cell nucleus we computed a total number of 356 features, including 12 dynamic features. The classes for cell nucleus classification were selected differently compared to Sect. 4.4, since here the focus of the analysis was on nuclear morphologies rather than on mitotic progression. We established thirteen classes, again including the seven cell cycle phases ((1) *interphase*, (2) *prophase*, (3) *prometaphase*, (4) *metaphase*, (5) *early anaphase* (*ana1*), (6) *late anaphase* (*ana2*), (7) *telophase*) as well as the following six nuclear morphology phenotypes: (8) *segregation problems* (mitotic nuclei with chromosome bridges), (9) *cell death*, (10) *binucleated* cells, (11) *multinucleated* cells, (12) nuclei with attached *micronuclei*, and (13) *lobulations* (abnormal nucleus shape characterized by concavities) (see Fig. 4.18). The training set for the final classifier included 24626 samples from 18 different image sequences which were manually annotated by experts using our *ImageJ*-based annotation tool (see Sect. 3.8.3). The number of training samples per class are given in Fig. 4.18. For classification we applied weighted support vector machines (see Sect. 3.6.2) where the class weights were computed by dividing the largest sample number (here interphase) by each class sample number (as proposed, e.g., in [179]). We did not include the final step of phase sequence parsing, since cell cycle phase durations were not the main interest of the study. The desired biological readout was the temporal distribution of morphological phenotypes based on cell populations.

To accelerate the computation, the original data set was split into multiple independent subsets and the described analysis steps were performed in parallel on the data subsets using multiple computers. Thus, computation of the final result was sped up, but at the same time it was not necessary to adapt the algorithms for parallelized computing. We used up to seven computers with AMD Opteron processors (2.2 to 2.4 GHz), 64 bit architecture, and 3.9 to 7.8 GByte RAM. To manage the computation processes that is, e.g., to start processes, to iterate and create directory structures, or to compress and delete files, we created *Perl* scripts and *Linux shell* scripts. The resulting data volumes and file numbers of intermediate and final results, as well as the computation times, are given in Tab. 4.31 below.

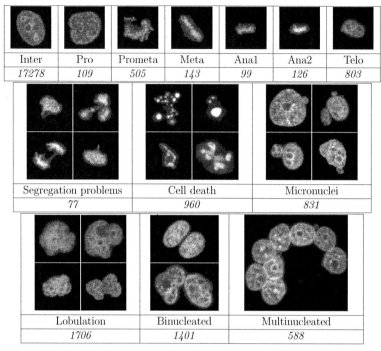

Inter	Pro	Prometa	Meta	Ana1	Ana2	Telo
17278	*109*	*505*	*143*	*99*	*126*	*803*

Segregation problems	Cell death	Micronuclei
77	*960*	*831*

Lobulation	Binucleated	Multinucleated
1706	*1401*	*588*

Figure 4.18: Example images for the 13 different classes, including seven normal cell cycle phases (first row) and six morphological phenotype classes (other rows). Below the class name the number of available training samples is given for each class.

To summarize the final results and automatically create comprehensive diagrams we used different scripts for *Perl*, *Gnuplot*, and *R*.

4.5.3   Experimental results

In the first step, we evaluated the classification performance of our system and optimized the composition of the training data set. To determine a reliable estimate of the classification performance we performed different cross validation experiments on the training set. Also, we compared the results for weighted and non-weighted support vector machines. After optimizing the setup for classification we trained the final classifier on the whole training set and applied it to the complete screening data. The overall analysis results were then further evaluated and comprehensive diagrams were generated to provide an overview of the resulting distribution of phenotypes.

Computational resources

Detailed analysis of a large high-throughput data set naturally also consumes a certain amount of computational resources. An overview of the here used resources is provided in Tab. 4.31 and will be discussed below. In this high-throughput screen we dealt with a source data volume of more than 250 GByte. In addition, we required around 220 GByte of storage space for intermediate results, such as segmentation and tracking results, extracted single cell images and image features, and the classifier output. Such results were stored to speed up potential follow-up studies. The disk space required per image sequence on average, as well as the total required disk space for the 951 image sequences is given in Tab. 4.31 (left). However, not only the data volume is of interest, but also the amount of generated files since large file numbers can slow down even simple processes in regular file systems. Here, particularly for the intermediate results large file numbers were produced (see Tab. 4.31, middle). Finally, the computation time is an important issue since even small differences in computation time of seconds or few minutes scale up to hours and days for very large data sets. The average computation time for one image sequence, as well as the total computation time for all image sequences is given on the right in Tab 4.31. In particular, the extraction of image features is very time consuming since a large number of 356 features was extracted per cell nucleus. However, this could be easily sped up by reducing the number of features if, on the other hand, one is willing to accept a decrease of the classification accuracies. Also hardware acceleration using graphical processing units (GPUs) and further parallelization of the computation provides significant speed up of the computation as we showed in a recent joint project [60, 61].

Classification accuracy

Data sampling for cross validation As already described in previous sections, cross validation is a useful technique to evaluate the classification accuracy. However, there are different strategies to split the data into cross validation sets which can lead to different results. In this application, for certain phenotype classes (e.g., lobulation and micronuclei) the morphology hardly changed over time, resulting in sequences of relatively similar single cell images. This was probably due to the fact that the respective knockdowns caused only relatively mild effects with which the treated cells could live on. In comparison, the morphological phenotypes in the above application (see Sect. 4.4) changed much more quickly over time. To yield a possibly high number of training samples, the sequences of cells showing the phenotype were annotated and then, each time step was used as single training sample. However,

	Data volume		File	Computation time	
	Avg./seq.	Total	number	Avg./seq.	Total
Segmentation	1.2 MB	1.1 GB	208143	6.9 min	109.0 h
Tracking	1.3 MB	1.2 GB	951	3.9 min	60.0 h
Single obj. extr.	48.9 MB	44.9 GB	11147409		
Feature extr.	8.6 MB	7.8 GB	951	74.4 min	1179.0 h
Classification	0.3 MB	0.3 GB	951	2.1 min	33.2 h
Total	60.3 MB	55.3 GB	11358405	87.3 min	1381.2 h

Table 4.31: Computational resources required for the analysis of a high-throughput screen broken down into the subsequent analysis steps (segmentation, tracking with extraction of single object stacks, feature extraction, and classification). From left to right the columns provide the average disk space per image sequence and the total disk space, the number of created files, and the average computation time per image sequence as well as the total computation time for each analysis step. Note that the images resulting from segmentation and single object stack extraction have been compressed (using Linux *gzip*). Note also that the time for training the classifier is not included.

applying *systematic sampling* (see, e.g., [84]), as done in the above applications, distributes these similar training samples evenly over the cross validation sets. This led to an overestimation of the classification accuracy since very similar samples were used for training and for testing. For our training data we yielded an overall classification accuracy of 96.4%. In particular, for almost all of the morphological phenotypes the accuracies were between 97% and 99% (see Appendix Tab. A-1.11). Also *random sampling* most probably would have resulted in a far too optimistic accuracy estimation since this sampling strategy also would have distributed the similar samples over the cross validation sets.

To provide a more realistic estimate we next split the data based on complete image sequences into cross validation sets, meaning that all samples from one image sequence were in the same cross validation set (we denote this strategy as *sequence-wise systematic sampling*). Since the training data was taken from 18 different image sequences we created nine cross validation sets to obtain equally sized sets, where each set included the data from two image sequences. However, even though the sets had approximately the same size, the data naturally was not stratified, i.e. the class probabilities were not equal for the different sets. Consequently, in several cross validation sets not all classes were represented, and for some classes training data was only available in one or two cross validation sets. Thus, in some runs of the cross validation there were no, or only very few training samples available for certain classes, which naturally led to an underestimation of the classification accuracy. With sequence-wise systematic sampling we yielded an overall classification accuracy of only 45.8% for this experiment (see Appendix Tab. A-1.11). A reasonable compromise between the two sampling extrema *systematic* and *sequence-wise systematic sampling* was found by *track-wise systematic sampling*. Here, whole trajectories were assigned to cross validation sets, with the result that all samples from one cell were in the same set, while cells from one sequence could be distributed over different sets. This way we could make sure that not too similar samples from one cell were used for training and testing, but also that each phenotype class was adequately represented in the training data. Using this sampling strategy we yielded overall accuracies of around 85%, more details are given in the following paragraph.

Weighted versus non-weighted SVMs   As already discussed in Sect. 4.4.3 we here again have a highly unbalanced training data set (see Fig. 4.18) with sample numbers between 77 (*segregation problems*) and 17278 (*interphase*). We applied weighted support vector machines to deal with this problem and yielded an overall classification accuracy of 85.0% (see Tab. 4.32(a)). In comparison, using non-weighted SVMs yielded an overall accuracy of 85.4% (see Tab. 4.32(b)) which again shows that both strategies yield comparable overall accuracies. However, weighted SVMs performed better for some of the classes with lower sample numbers, for example, for class *micronuclei* (increase from 17.1% to 19.4%). Since in this application the abnormal nuclear phenotypes were of highest interest, an accuracy decrease for the normal mitotic phases in favor of the abnormal phenotype classes was acceptable. For this reason we used weighted SVMs for classification of the complete screening data set. The final classifier was trained based on the entire training data set and the weights were computed by dividing the largest sample number (here interphase) by each class sample number (e.g., [179]).

Evaluation of the final classification results

To summarize the classification results of the whole data set and to provide a comprehensive overview, different types of diagrams were generated. First, for each image sequence one chart was generated visualizing the occurrence of nuclear morphology phenotypes throughout the image sequence. Therefore, at each time step of an image sequence the relative number of cells per class was plotted (for the nuclear morphology phenotypes and prometaphase as an indicator for mitosis events). In addition the diagrams show the normalized total number of cell nuclei for each time step. Example diagrams for one control and one treated experiment are given in

a												
True	**Classifier Output**											
Class	**1**	**2**	**3**	**4**	**5**	**6**	**7**	**8**	**9**	**10**	**11**	**12**	**13**
1	**16567**	6	6	13	0	0	438	0	0	78	5	52	113
2	21	**71**	11	0	1	0	0	0	0	1	3	0	1
3	6	7	**362**	18	1	1	4	1	71	1	0	31	2
4	2	0	34	**86**	1	3	7	0	9	0	0	0	1
5	1	0	7	3	**70**	14	0	0	3	0	1	0	0
6	0	0	1	3	8	**87**	18	1	7	0	1	0	0
7	105	0	4	7	0	19	**639**	21	7	0	0	0	1
8	1	0	5	1	2	1	17	**44**	4	1	1	0	0
9	2	0	88	11	5	9	45	0	**758**	0	42	0	0
10	109	0	0	0	0	0	1	0	8	**1030**	218	0	35
11	9	0	0	0	0	0	0	0	91	216	**272**	0	0
12	531	23	1	0	0	0	69	0	0	0	11	**161**	35
13	669	0	0	0	0	0	28	0	0	17	141	58	**793**
Acc. [%]	95.9	65.1	71.7	60.1	70.7	69.1	79.6	57.1	79.0	73.5	46.3	19.4	46.5

b												
True	**Classifier Output**											
Class	**1**	**2**	**3**	**4**	**5**	**6**	**7**	**8**	**9**	**10**	**11**	**12**	**13**
1	**16659**	7	2	13	1	0	388	0	0	74	2	42	90
2	21	**71**	11	0	1	0	0	0	0	1	3	0	1
3	6	7	**362**	19	1	1	3	1	71	1	0	31	2
4	1	0	29	**93**	1	3	6	0	9	0	0	0	1
5	1	0	7	3	**69**	14	0	1	3	0	1	0	0
6	0	0	1	4	8	**87**	18	1	6	0	1	0	0
7	123	0	4	7	0	24	**615**	19	7	0	0	0	4
8	2	0	5	1	2	1	16	**44**	5	1	0	0	0
9	1	1	88	11	5	9	45	0	**758**	0	42	0	0
10	88	0	0	0	0	0	1	0	8	**1069**	201	0	34
11	7	0	0	0	0	0	0	0	91	218	**272**	0	0
12	569	23	1	0	0	0	49	0	0	0	11	**142**	36
13	639	0	0	0	0	0	32	0	0	26	141	70	**798**
Acc. [%]	96.4	65.1	71.7	65.0	69.7	69.1	76.6	57.1	79.0	76.3	46.3	17.1	46.8

Table 4.32: Confusion matrix for SVM classification using five-fold cross validation on the training set. (a) *Weighted* SVM with an overall accuracy of 85.0%, (b) *nonweighted* SVM with an overall accuracy of 85.4%. The classes corresponding to the numbers are: (1) interphase, (2) prophase, (3) prometaphase, (4) metaphase, (5) early anaphase, (6) late anaphase, (7) telophase, (8) segregation problems, (9) cell death, (10) binucleated, (11) multinucleated, (12) micronuclei, and (13) lobulation. The last row gives the classification accuracies per class in percent.

Fig. 4.19. Second, the results of corresponding replicate experiments were combined. To this end, for each group of replicate experiments (usually three) the occurring class samples were added up per time step. Again, the temporal distribution of classes was plotted as described above for the combination of the replicates to provide a more reliable impression of the phenotype development. Third, the class frequencies were summed up over all time steps and plotted for each combined experiment in form of a heatmap, where the rows represent the different experiments and the columns the phenotype classes. The rows of the heatmap were standardized to a mean value of zero and a standard deviation of one to make the entries better comparable. Figure 4.20 shows an example heatmap for a group of 70 experiments. Note that in total five heatmaps of this type were generated for the complete screen. In the heatmaps we can identify at one glance experiments causing an increased occurrence of any phenotype class. This allows the biologists to choose the most interesting experiments from the heatmap and consider the respective temporally resolved plots to study phenotype development. Finally, only for the most interesting cases they need to view the original image sequences.

4.5.4 Summary and conclusion

In this section, we presented results for fully automatic analysis of a high-content screen of 104 nuclear envelope proteins. We applied our extended approaches for segmentation and tracking (see Sect. 4.4), as well as feature extraction and classification to 951 multi-cell 3D image sequences with nuclear morphology phenotypes. First, we systematically evaluated the classification accuracy using cross validation with different sampling strategies (systematic sampling, sequence-wise systematic sampling, and track-wise systematic sampling). We identified *track-wise* systematic sampling as most adequate for this application since training samples within one trajectory often were extremely similar. Here, we used weighted support vector machines to deal with the unbalanced training data set which resulted in an overall accuracy of 85.0% for classification into 13 classes. Comparing the results with the output of non-weighted support vector machines showed that although the non-weighted scheme produced a comparable overall accuracy, the weighted scheme was still more useful in our application since the classes with low sample numbers were classified more accurately.

Our approach was finally applied to automatically analyze the whole screening data set. The computation was performed on multiple computers in parallel. To provide an estimate of the computational requirements for the evaluation of such

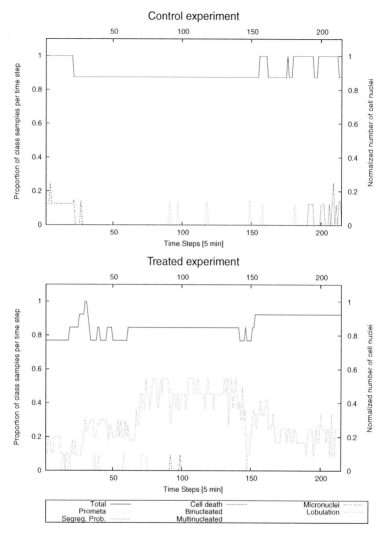

Figure 4.19: Example diagrams of the relative phenotype frequency over time. The absolute class frequencies are normalized with the total cell number at each time step. The red solid line shows the total cell number per time step normalized with the maximum cell number of the sequence. The displayed classes are (3) prometaphase (green dashed) to represent mitotic events, and the nuclear morphology phenotypes (8) segregation problems (blue dashed), (9) cell death (pink dashed), (10) binucleated (turquoise dashed), (11) multinucleated (yellow dashed), (12) micronuclei (black dashed), and (13) lobulation (orange dashed).

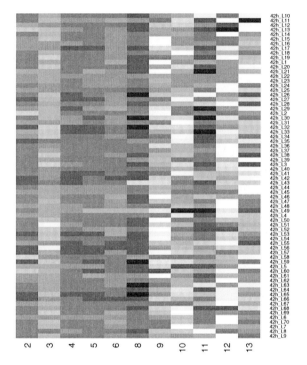

Figure 4.20: Example heatmap showing the class frequencies for the classes (2) prophase, (3) prometaphase, (4) metaphase, (5) early anaphase, (6) late anaphase, (8) segregation problems, (9) cell death, (10) binucleated, (11) multinucleated, (12) micronuclei, and (13) lobulation as columns, and 70 experiments (each including three replicates) as rows. Each row was standardized to a mean value of zero and a standard deviation of one. Yellow denotes high frequency, red low frequency of occurrence.

a high-content screen we presented an overview of the consumed computational resources. The analysis results were used to generate comprehensive diagrams for each image sequence, as well as for the combined data of different replicate experiments. Finally, heatmaps of the phenotype prevalence were created which provide a good overview of all experiments.

Chapter 5

Summary and Future Work

In this thesis, we addressed the problem of automatic analysis of multi-dimensional live cell images in the context of high-throughput high-content screens. Increasingly automated laboratories allow generating large amounts of image data within relatively short time frames, however, adequate methods to analyze the resulting data are still often lacking. Depending on the desired analysis readout as well as on the image data characteristics, complex image analysis workflows are required for automatic analysis. Particularly, for the question of detailed cell cycle analysis only few approaches have been developed so far.

5.1 Summary

In this thesis, we developed an automatic approach for cell cycle analysis based on 3D multi-cell image sequences. Our approach allows studying nuclear phenotypes over time on a single cell basis and thus enables detailed cell cycle analysis. In particular, our approach allows accurate determination of mitotic phase durations. Below, we summarize the contributions of this thesis:

- We developed a complex *image analysis workflow* for nuclear phenotype and cell cycle analysis, comprising robust methods for segmentation, tracking, feature extraction, classification, and phase sequence parsing. Our approach allows high-throughput analysis of 2D and 3D image sequences by reducing the 3D image stacks to 2D, based on cell-wise most informative slice selection. We have applied our approach successfully to more than 1000 3D image sequences including in total more than one million 2D images from four different screening data sets. In particular, we demonstrated the ability of our approach to accurately measure mitotic phase durations in a poof-of-principle experiment.

- We developed a robust approach for cell nucleus *segmentation* which can cope with very difficult images of abnormal nuclear phenotypes. The basic algorithm is based on region-adaptive thresholding with overlapping regions. We validated the region-adaptive thresholding algorithm on a set of real images and showed that it performs better than schemes using non-overlapping regions. We extended our approach to enable segmentation of dim attached micronuclei by masking the bright main nucleus. Furthermore, detached nucleus parts, which normally are segmented as single objects, are detected and merged to the main nucleus. The segmentation accuracy for the extended method was systematically evaluated based on real data, and it turned out that we yield generally high segmentation accuracies. Also, we compared the performance of our approach with two other segmentation approaches commonly used for cell nucleus segmentation and showed that our approach performs significantly better on our images.

- For tracking of mitotic cell nuclei we developed a feature point tracking scheme that combines correspondence finding and mitosis detection. Correspondences are determined based on Euclidean distance and a smoothness cost function, while for mitosis detection we developed a new *mitosis detection measure*. Because of the temporal resolution of several minutes in our application, the separation of daughter cell nuclei during mitosis is represented by only a few images, and thus, appears as a sudden duplication of one object into two distant objects. Therefore, identification of mitosis is relatively difficult. Our new mitosis detection method relies on morphological and topological features of mother and daughter cell nuclei (size, mean intensity, and Euclidean distance). These features are evaluated in relation to the population averages of nucleus size and intensity instead of predefined thresholds. Therefore, only a small number of parameters to weight the importance of the different features have to be selected by the user. In comparison to the overlap distance ratio as a frequently used measure for mitosis detection our new criterion performs significantly better. We evaluated the tracking and the mitosis detection performance based on real image sequences and yielded high accuracies. Moreover, we compared the performance of our tracking scheme with two other approaches for cell tracking. We found that our approach yielded a slightly higher accuracy than both approaches, and a main advantage of our approach is that it allows accurate mitosis detection.

- In addition to *static* image features used to characterize cell nuclei, we developed *dynamic features* representing morphological changes between consecutive time steps. Based on the tracking result these features include information from the previous and subsequent image frames. We showed that using both static and dynamic features significantly increases the classification accuracies, particularly for classes with high variability. Moreover, we tested and compared different strategies for feature extraction from 3D image stacks, and showed that extracting features from maximum intensity projection images as well as from the most informative slice allows more accurate classification than using features from only one of the two different image types.

Furthermore, we compared different feature reduction methods and different classification schemes based on real image data. It turned out that support vector machines performed best for our application. For the here considered application we built accurate classifiers for cell nucleus classification into four to thirteen classes, including regular cell cycle phases as well as abnormal nuclear phenotypes and apoptosis.

- For robust determination of cell cycle phase durations we developed an error-correcting *phase sequence parser* based on a state model. Our method checks the consistency of phase sequences, corrects errors, and accurately determines cell cycle phase durations automatically. To our knowledge a comparable method for detailed phase sequence analysis and quantification of phase durations has not been described earlier. Based on the resulting phase durations we performed detailed analysis of temporal phenotypes such as phase prolongations or shortenings. Moreover, we investigated correlations between temporal phenotypes and the resulting morphological phenotypes (e.g., detached chromosomes or micronuclei) based on single-cell analysis. In a proof-of-principle experiment we showed that our approach was able to accurately detect cell cycle phase delays and correlate them with morphological phenotypes.

- Our approach was successfully applied for the evaluation of a large number of live cell images and the performance was analyzed systematically. In particular, we analyzed a full RNAi knockdown screen targeting 104 genes. To prove that our approach is also directly applicable to images acquired with different experimental settings, we applied it to images of a different cell line and images acquired with a different fluorescence microscopy platform. It turned out that we yielded comparably high accuracies as for the previously analyzed images by adapting only a few parameter values.

- We also developed user-friendly *software tools* for the annotation of ground truth data and for visualization of the tracking and classification results. Correctly annotated training data is essential for automatic classification, and in most applications considerably more than 100 samples per class were annotated manually. In particular, to evaluate the performance of the phase length determination several complete image sequences had to be annotated to provide ground truth. Consequently, an ergonomic interface which facilitates data annotation and minimizes annotation errors is important and can increase the overall accuracy of the results. Also, a good visualization of the results is required, in particular to communicate the results between researchers of different research areas. We implemented the respective tools as plugins for *ImageJ* which allows easy installation on different computing platforms.

In summary, the work presented in this thesis advances automatic image analysis in the context of high-throughput high-content screening. Automatic analysis of temporal mitotic phenotypes as presented here provides new possibilities for large-scale cell cycle analysis. Since our approach is composed from independent modules it can be easily modified and adapted for the analysis of related problems. We also proved that our approach can be directly applied to images acquired with different experimental settings and still yields high accuracies.

5.2   Future Work

In the following, we discuss a list of issues that could further improve our approach and should be addressed in future work.

- Fluorescence microscopy allows not only acquiring the emitted fluorescence signals, but also the transmission light can be captured in a separate channel. This data is readily available and does not require any staining or additional experimental work. Since the quality of the transmission light images is usually very low, they are often not acquired although they might provide additional information (in our application they were provided only for some cases so we could not make systematic use of them). However, in future work, methods to explore the transmission light channel could be studied. Note that the contours of mitotic cells usually can be well identified based on the transmission light images, and this information could be used to improve the segmentation result. Moreover, mitosis detection during tracking could be probably improved, and also for classification this information could be used as an additional feature.

- Concerning the computation time it turned out that feature extraction is by far the most time consuming step of the overall analysis. Extraction of the full set of more than 350 features on average takes about six times longer than all the other processing steps together. Therefore, the highest potential to accelerate the analysis process lies in accelerating the feature extraction. To this end, future work should include further soft- and hardware acceleration of the feature computation as it has already been done in a related project for the Haralick texture features [60, 61]. Also, further techniques for feature selection could be studied to reduce the overall number of features without significantly decreasing the classification accuracy.

- To improve classification, information on previous and subsequent time steps could be exploited to a greater extent, e.g., by combining the classifier and the state model for phase sequence parsing. In particular, we could use probabilistic state models such as hidden Markov models or other context-based mixture models for sequence classification. This integration of model- and feature-based classification could increase the overall accuracy and robustness of our approach. However, the drawback of using hidden Markov models for classification and phase sequence analysis is that for training they require sequences of samples providing sufficiently high numbers of all possible phase transitions. This means an enormous effort to manually annotate all the training data sequences, given the large number of classes and transitions we considered here. Moreover, if a conventional HMM is trained based on control sequences (which usually do not show any phase prolongations or shortenings), the learned phase transition probabilities will be less suited for accurate classification of sequences with significantly prolonged or shortened phases. Thus, adaptations would be necessary to enable accurate phase length determination for sequences with normal and abnormal mitotic progression.

- Up to now our finite state machine for phase sequence parsing implements all phases and phase transitions statically. Consequently, for applications with different classes and different state transitions the software implementation has to be adapted. To make our scheme more flexible and directly usable for other applications with different classes and state transitions, the FSM implementation should be extended to allow dynamic instantiation of specific finite state machines based on given transition matrices.

- For high-throughput analysis of large data sets in our case a huge number of relatively small files (e.g., single cell images and feature files) are generated and stored as intermediate results. Using a regular network file system for data management this can lead to a significant slowdown of input/output processes. To accelerate the analysis process one could instead operate on an optimized virtual file system that allows faster file access, e.g., by reducing the maintenance of file metadata.

- Finally, we note that our approach, in particular, the segmentation approach is adapted and optimized for fluorescence microscopy images of cell nuclei. Thus, for analyzing images from a very different type of microscope (e.g., phase contrast microscope) it might be advantageous to plug in a better adapted segmentation approach. Also the mitosis detection measure relies on features which are specific for fluorescence microscopy images of cell nuclei. With our approach it is possible to exchange analysis steps or include additional analysis steps into the workflow if necessary. For example, to deal with spatial positioning issues that may occur in live cell imaging, a registration step can be included into the analysis pipeline, as we have done in a related study [100]. Consequently, in principle our approach can be adapted to different image types and applications by exchanging or adding single modules in the analysis pipeline.

Appendix

A-1   Additional Tables

True Class	Classifier Output											
	1	**2**	**3**	**4**	**5**	**6**	**7**	**8**	**9**	**10**	**11**	**12**
1	**19442**	29	1	0	0	0	153	303	0	0	0	3
2	211	**409**	19	1	0	0	3	8	0	0	0	0
3	8	20	**5251**	90	3	3	2	11	2	2	3	19
4	8	1	94	**576**	2	2	7	2	0	1	2	5
5	0	0	16	12	**76**	6	0	0	0	3	1	0
6	3	0	2	1	3	**257**	11	0	0	0	6	1
7	212	0	2	4	0	4	**1451**	23	0	0	1	48
8	929	4	20	1	0	0	52	**7434**	0	0	0	65
9	5	0	24	0	0	0	0	1	**217**	0	1	0
10	0	0	23	1	9	0	1	0	0	**10**	0	2
11	2	0	24	2	2	17	6	0	0	1	**21**	2
12	27	0	36	4	0	1	167	155	0	0	3	**437**
Acc. [%]	97.6	62.8	97.0	82.3	66.7	90.5	83.2	87.4	87.5	21.7	27.3	52.7

Table A-1.1: Confusion matrix using a feature set with 100 independent components. Here, a combined classifier was used including data from the nocodazole and RNAi experiments. Overall classification accuracy: 92.3%.

		Nocodazole				RNAi	
		Control	Low	Medium	High	Control	Treated
Man	n	57	44	68	120	27	29
	p-value	$2.0 \cdot 10^{-10}$	$5.3 \cdot 10^{-4}$	$1.1 \cdot 10^{-10}$	$9.2 \cdot 10^{-11}$	$9.7 \cdot 10^{-8}$	0.06
Auto	n	180	37	21	45	66	115
	p-value	$1.8 \cdot 10^{-15}$	$8.6 \cdot 10^{-5}$	$1.8 \cdot 10^{-7}$	$1.5 \cdot 10^{-7}$	$5.7 \cdot 10^{-11}$	$2.0 \cdot 10^{-10}$

Table A-1.2: Results of Shapiro-Wilk normality test on the prometaphase length distributions of manually and automatically annotated data for the nocodazole and RNAi experiments; n is the number of samples.

a		Manual			Automatic		
		Control-Low	Control-Medium	Control-High	Control-Low	Control-Medium	Control-High
Shortened	Interphase						
	n_1,n_2	184, 142	184, 172	184, 142	545, 70	545, 48	545, 65
	p-value	$9.8 \cdot 10^{-3}$	$1.3 \cdot 10^{-10}$	$3.8 \cdot 10^{-3}$	$5.1 \cdot 10^{-3}$	$9.6 \cdot 10^{-3}$	0.88
	Telophase						
	n_1,n_2	104, 52	104, 113	104, 52	358, 44	358, 40	358, 45
	p-value	$1.9 \cdot 10^{-4}$	$1.3 \cdot 10^{-4}$	$3.5 \cdot 10^{-6}$	$9.9 \cdot 10^{-5}$	$4.8 \cdot 10^{-3}$	0.05
	Abnormal interphase						
	n_1,n_2	5, 24	5, 36	5, 157	55, 27	55, 18	55, 24
	p-value	$6.6 \cdot 10^{-4}$	0.04	$3.9 \cdot 10^{-3}$	0.94	0.94	0.55
Prolonged	Late anaphase						
	n_1,n_2	95, 38	95, 72	95, 19	262, 18	262, 24	262, 31
	p-value	0.01	NA	0.01	0.24	0.07	0.42
	Abnormal Telophase						
	n_1,n_2	5, 28	5, 40	5, 87	99, 27	99, 18	99, 30
	p-value	0.04	0.13	$8.4 \cdot 10^{-3}$	0.59	0.46	0.02

b			Man	Auto
			Control-Treated	Control-Treated
Shortened	Telophase	n_1,n_2	51, 9	90, 54
		p-value	0.95	0.03
	Abnormal interphase	n_1,n_2	14, 14	27, 32
		p-value	0.14	$2.7 \cdot 10^{-6}$
Prolonged	Early anaphase	n_1,n_2	19, 5	24, 22
		p-value	NA	0.02
	Late anaphase	n_1,n_2	48, 9	62, 27
		p-value	NA	0.02
	Abnormal telophase	n_1,n_2	3, 6	13, 35
		p-value	0.01	0.29

Table A-1.3: Results of Mann-Whitney U tests for all phases, (a) for the nocodazole experiments, and (b) for the RNAi experiments. One-sided tests were performed with alternative hypothesis "true shift is greater/smaller than 0"for prolonged and shortened phases, respectively. n_1, n_2 are the sample numbers, NA resulted in case all samples were identical.

Man-Auto	Nocodazole				RNAi	
	Control	Low	Medium	High	Control	Treated
n_1, n_2	57, 180	44, 37	68, 21	120, 45	27, 66	29, 115
p-value	0.04	0.06	0.92	0.28	0.25	0.02

Table A-1.4: Results of Mann-Whitney U tests on the prometaphase length distributions of manually annotated data against the automatically annotated data for nocodazole and RNAi experiments. Two-sided test were performed with alternative hypothesis "true shift is not equal to 0", n_1, n_2 are the sample numbers of the manually and automatically annotated data, respectively.

	Nocodazole		RNAi	
	Man	Auto	Man	Auto
Mean length μ	2.00	2.43	2.04	2.94
Stddev of lengths σ	1.24	1.51	0.81	2.61
Significance threshold $(\mu + 2\sigma)$	4.48	5.46	3.65	8.17
Threshold in minutes	31 min	38 min	26 min	57 min

Table A-1.5: Mean values and standard deviations of prometaphase lengths determined for the control experiments (in time steps), and significance thresholds to determine prolonged prometaphases (in time steps and minutes).

		Including cases with $r = 0$				Excluding cases with $r = 0$			
		Nocodazole		RNAi		Nocodazole		RNAi	
		Man	Auto	Man	Auto	Man	Auto	Man	Auto
Normal	Low	0.23	0.27	0.17	0.30	0.41	0.41	0.90	0.68
	Medium	0.32	0.31			0.49	0.36		
	High	0.35	0.25			0.69	0.41		
Prolonged	Low	0.16	0.17	0.04	0.15	0.31	0.44	0.20	0.40
	Medium	0.09	0.31			0.27	0.38		
	High	0.14	0.11			0.51	0.30		
Control		0.02	0.12	0.07	0.11	0.25	0.24	0.19	0.30

Table A-1.6: Mean values for ratio r (number of post-prometaphase steps with abnormal phenotypes divided by the total number of post-prometaphase time steps) for all experiments. In the left part, all cases were used for computing the mean (including cases with a ratio of $r = 0$), in the right part, only samples with a ratio of $r > 0$ were considered.

		Nocodazole		RNAi	
		Man	Auto	Man	Auto
n		272	261	51	171
Kendall	τ	0.33	0.24	-0.03	0.11
	p-value	$5.5 \cdot 10^{-13}$	$2.2 \cdot 10^{-7}$	0.39	0.03
Spearman	ρ	0.41	0.31	-0.04	0.13
	p-value	$8.2 \cdot 10^{-13}$	$1.7 \cdot 10^{-7}$	0.40	0.05

Table A-1.7: Correlation coefficients τ and ρ for prometaphase duration and ratio r (probability of following abnormal morphologies), including cases with ratio $r = 0$. Additionally, the p-values for the significance tests and numbers of measurements n are given.

	Total no. of cells	Correct	Under-segmented	Over-segmented	Partly/not segmented	Accuracy [%]
a	4008	3981	27	0	0	**99.3**
b	9226	9206	5	0	15	**99.8**

Table A-1.8: Evaluation of the segmentation accuracy for (a) NRK (normal rat kidney) cells acquired with an LSM 510 Meta point-scanning microscope (six image sequences), and (b) HeLa cells acquired with an LSM 5 LIVE line-scanning microscope (four image sequences).

	No. of tracks	No. of links	Errors segm.	Errors mitosis detect.	Errors corresp. finding	Linking accur. [%]	Mitoses found	Mitosis detect. acc. [%]
a	66	4073	5	2	6	99.7	14/16	**87.5**
b	32	8431	4	0	0	99.9	22/22	**100.0**

Table A-1.9: Evaluation of the tracking accuracy for (a) NRK (normal rat kidney) cells acquired with an LSM 510 Meta point-scanning microscope (six image sequences), and (b) HeLa cells acquired with an LSM 5 LIVE line-scanning microscope (four image sequences).

True Class	Classifier Output							Accur.
a								
	Inter	Pro	Prom.	Meta	Ana1	Ana2	Telo	[%]
Inter	**3020**	30	0	0	0	0	82	**96.4**
Pro	18	**51**	3	0	0	0	1	**69.9**
Prom.	3	3	**34**	1	1	0	1	**79.1**
Meta	2	1	8	**9**	0	0	2	**40.9**
Ana1	0	0	1	0	**26**	6	0	**74.3**
Ana2	0	0	0	1	5	**27**	9	**64.3**
Telo	63	0	0	1	0	7	**135**	**65.5**

True Class	Classifier Output							Accur.
b								
	Inter	Pro	Prom.	Meta	Ana1	Ana2	Telo	[%]
Inter	**7724**	1	0	0	0	0	74	**99.0**
Pro	7	**41**	0	0	0	0	0	**85.4**
Prom.	1	2	**58**	5	0	0	0	**87.9**
Meta	1	0	7	**56**	2	0	0	**84.8**
Ana1	1	0	0	3	**19**	5	2	**63.3**
Ana2	2	0	0	0	2	**44**	10	**75.9**
Telo	53	0	0	0	0	5	**242**	**80.7**

Table A-1.10: Confusion matrices for classification of (**a**) NRK (normal rat kidney) cells acquired with an LSM 510 Meta point-scanning microscope (six image sequences), and (**b**) HeLa cells acquired with an LSM 5 LIVE line-scanning microscope (four image sequences).

a

True Class	Classifier Output													
	1	2	3	4	5	6	7	8	9	10	11	12	13	14
1	**1824**	1	2	0	0	0	21	0	5	3	0	9	8	2
2	9	**88**	8	0	1	0	0	0	2	2	0	3	0	0
3	4	9	**459**	15	4	0	2	0	9	3	0	3	1	0
4	1	0	17	**120**	1	1	3	0	1	0	0	1	0	0
5	2	0	6	1	**87**	7	2	0	2	0	0	0	0	3
6	1	0	0	1	9	**94**	21	1	4	0	0	0	0	1
7	11	0	2	2	0	13	**777**	3	0	0	0	0	8	1
8	0	0	4	1	1	3	9	**48**	1	2	0	0	0	0
9	5	0	4	3	4	3	0	0	**959**	3	0	4	0	3
10	2	0	1	0	0	0	0	0	4	**1406**	1	2	0	1
11	0	0	0	0	0	0	0	0	4	1	**588**	4	0	0
12	5	0	2	0	0	0	2	0	3	2	4	**820**	0	0
13	2	0	0	0	0	0	3	0	2	2	0	2	**1624**	0
14	1	0	0	0	0	0	0	0	1	0	3	0	0	**740**
Acc. [%]	97	78	90	83	79	71	95	70	97	99	99	98	99	99

b

True Class	Classifier Output													
	1	2	3	4	5	6	7	8	9	10	11	12	13	14
1	**1289**	2	10	4	3	14	154	1	1	20	0	51	304	22
2	8	**72**	21	0	4	1	0	0	0	1	1	2	2	1
3	0	20	**299**	40	7	16	7	1	112	1	0	3	3	0
4	2	0	32	**75**	3	19	1	1	8	0	3	1	0	0
5	5	0	8	3	**70**	17	0	0	4	0	3	0	0	0
6	2	0	3	8	16	**73**	17	1	6	0	6	0	0	0
7	54	0	14	14	1	74	**616**	9	24	0	1	5	5	0
8	0	0	3	4	0	3	55	**1**	0	1	0	1	1	0
9	16	2	105	46	24	42	23	1	**670**	2	56	0	1	0
10	17	1	8	0	0	0	1	1	0	**1125**	151	0	99	14
11	2	0	1	0	0	0	0	0	94	187	**206**	0	0	107
12	534	2	20	0	0	0	164	0	7	38	2	**5**	66	0
13	1127	96	3	0	0	0	41	0	0	72	0	146	**75**	75
14	258	1	1	0	0	0	0	0	0	0	484	0	1	**0**
Acc. [%]	69	64	59	52	64	55	75	0	68	79	35	0	1	0

Table A-1.11: Confusion matrix for SVM classification using cross validation with different sampling strategies. (**a**) *Systematic sampling*, overall accuracy 96.4% using five-fold cross validation, (**b**) *sequence-wise sampling*, overall accuracy 45.8% using nine-fold cross validation. The corresponding classes are: (1) interphase, (2) prophase, (3) prometaphase, (4) metaphase, (5) early anaphase, (6) late anaphase, (7) telophase, (8) segregation problems, (9) cell death, (10) binucleated, (11) multinucleated, (12) micronuclei, and (13) lobulation. Note that here a 14th class 'Large nuclei' was considered which was not used in the final evaluation and a reduced number of interphases was used. The last row gives the classification accuracies per class in percent.

A-2 Additional Figures

Abnormal
interphase

Abnormal
early
anaphase

Abnormal
late
anaphase

Abnormal
telophase

Cell
death

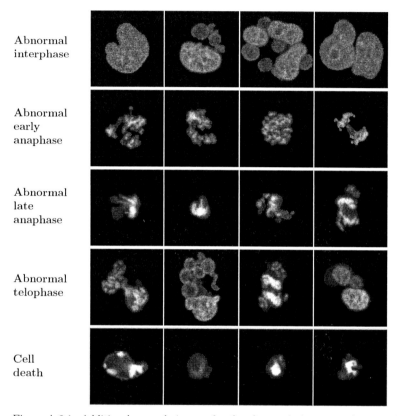

Figure A-2.1: Additional example images for the abnormal phenotype classes and class cell death caused by treatment with the spindle poison nocodazole. The images were taken from the training data set.

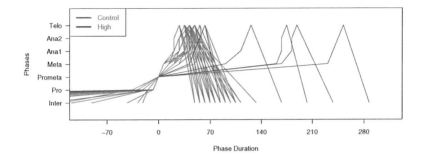

Figure A-2.2: Example phase sequences for single cells based on automated annotation of control and nocodazole treated cells. The phase sequences are temporally
aligned at the transition between prophase and prometaphase. x-axis: phase length
in minutes.

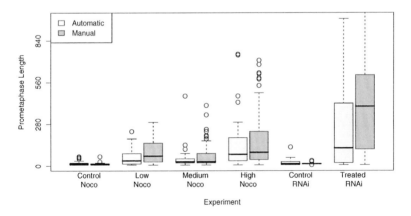

Figure A-2.3: Boxplots of prometaphase lengths for all control sequences and all
sequences treated with different concentrations of nocodazole and ch-TOG siRNA.
The yellow boxes represent automatically annotated data, the orange boxes manually annotated data. The upper border of each box represents the 75% quantile, the
bold line within the box the median, and the lower border of the box represents the
25% quantile. The whiskers reach at maximum 1.5 times the interquantile range,
the circles are outliers above or below the range of the whiskers.

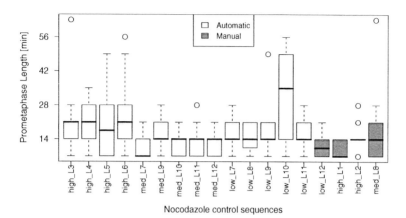

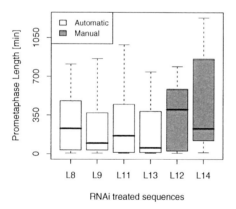

Figure A-2.4: Prometaphase durations for the image sequences of the nocodazole control experiments (top), and the RNAi treated experiments (bottom). The automatically annotated sequences are displayed in light gray, the manually annotated sequences in dark gray. The upper border of each box represents the 75% quantile, the bold line within the box the median, and the lower border of the box represents the 25% quantile. The whiskers reach at maximum 1.5 times the interquantile range, the circles are outliers above or below the range of the whiskers.

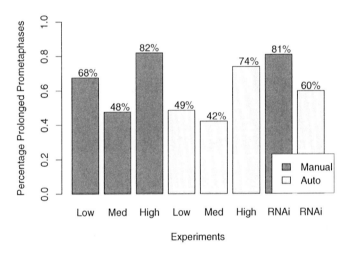

Figure A-2.5: Percentage of prolonged prometaphases in the treated experiments, determined using the significance thresholds as given in Appendix Tab. A-1.5.

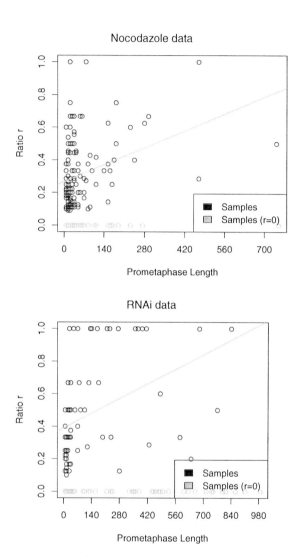

Figure A-2.6: Scatter plots of prometaphase duration versus ratio r for the automatically annotated nocodazole (top) and RNAi (bottom) data. After excluding samples with r = 0 (orange points) a regression line (green) was fit to the data. R^2 (the fraction of variance explained by the model) is 0.13 for the nocodazole data and 0.17 for the RNAi data.

Bibliography

[1] Apache Xerces XML parser. Software available at http://xerces.apache.org/.

[2] CellProfiler project. Software available at http://www.cellprofiler.org/.

[3] Extensible Markup Language XML. Specification available at http://www.w3.org/XML/.

[4] MineIt. Platform for data mining, developed at the division Theoretical Bioinformatics, DKFZ Heidelberg, http://www.dkfz.de/tbi/.

[5] MitoCheck EU project. Project homepage at http://www.mitocheck.org/.

[6] U. Adiga and B.B. Chaudhuri. An efficient method based on watershed and rule-based merging for segmentation of 3-D histo-pathological images. *Pattern Recognition*, 34(7):1449–1458, 2001.

[7] U. Adiga, R. Malladi, R. Fernandez-Gonzalez, and C. Ortiz de Solórzano. High-throughput analysis of multispectral images of breast cancer tissue. *IEEE Transactions on Image Processing*, 15(8):2259–2268, Aug 2006.

[8] A.V. Aho, R. Sethi, and J.D. Ullman. *Compilers: Principles, Techniques and Tools*. Addison-Wesley, Reading, MA, USA, 1986.

[9] B. Alberts, A. Johnson, J. Lewis, M. Raff, K. Roberts, and P. Walter. *Molecular Biology of the Cell*. Garland Science, New York, NY, USA, 5th edition, Dec 2007.

[10] F. Albregtsen. Non-parametric histogram thresholding methods – error versus relative object area. In K. A. H. Høgda, B. Braathen, and K. Heia, editors, *Proc. 8th Scadinavian Conf. Image Analysis*, pages 273–280, Tromsø, Norway, May 25–28 1993.

[11] W. Bai, X. Zhou, J. Zhu, L. Ji, and S.T.C. Wong. Tracking of migrating glioma cells in feature space. In J. Fessler and T. Denney, editors, *Proc.*

IEEE Internat. Symposium on Biomedical Imaging: From Nano to Macro (ISBI'2007), pages 272–275, Arlington, VA, USA, Apr 12–15 2007.

[12] W. Beaver, D. Kosman, G. Tedeschi, E. Bier, W. McGinnis, and Y. Freund. Segmentation of nuclei in confocal image stacks using performance based thresholding. In J. Fessler and T. Denney, editors, *Proc. IEEE Internat. Symposium on Biomedical Imaging: From Nano to Macro (ISBI'2007)*, pages 1044–1047, Arlington, VA, USA, Apr 12–15 2007.

[13] M. Belson, A.W. Dudley, and R.S. Ledley. Automatic computer measurements of neurons. *Pattern Recognition*, 1(2):119–128, 1968.

[14] C.M. Bishop. *Neural networks for pattern recognition.* Oxford University Press, New York, 2000.

[15] H.A.P. Blom. An efficient filter for abruptly changing systems. In *Proc. 23rd IEEE Conference on Decision and Control*, volume 23, pages 656–658, Dec 1984.

[16] M.V. Boland, M.K. Markey, and R.F. Murphy. Automated recognition of patterns characteristic of subcellular structures in fluorescence microscopy images. *Cytometry*, 33:366–375, 1998.

[17] M.V. Boland and R.F. Murphy. A neural network classifier capable of recognizing the patterns of all major subcellular structures in fluorescence microscope images of HeLa cells. *Bioinformatics*, 17(12):1213–1223, 2001.

[18] T.L. Booth. *Sequential Machines and Automata Theory.* John Wiley and Sons, New York, NY, USA, 1st edition, 1967.

[19] L. Bottou, C. Cortes, J.S. Denker, H. Drucker, I. Guyon, L.D. Jackel, Y. Le-Cun, U.A. Muller, E. Sackinger, P. Simard, and V. Vapnik. Comparison of classifier methods: a case study in handwritten digit recognition. In *Proc. 12th Internat. Conf. on Pattern Recognition*, pages 77–87, Jerusalem, Israel, 1994. IEEE Computer Society Press.

[20] L. Breiman. Random forests. *Machine Learning*, 45(1):5–32, 2001.

[21] L. Breiman, J.H. Friedman, R.A. Olshen, and C.J. Stone. *Classification and regression trees.* Chapman and Hall, New York, 1984.

[22] F. Bunyak, K. Palaniappan, S.K. Nath, T.I. Baskin, and G. Dong. Quantitative cell motility for in vitro wound healing using level set-based active

contour tracking. In J. Kovačević and E. Meijering, editors, *Proc. IEEE Internat. Symposium on Biomedical Imaging: From Nano to Macro (ISBI'2006)*, pages 1040–1043, Arlington, VA, USA, Apr 6–9 2006.

[23] C.J.C. Burges. A tutorial on support vector machines for pattern recognition. *Data Mining and Knowledge Discovery*, 2:121–167, 1998.

[24] H. Burkhardt and S. Siggelkow. *Invariant features in pattern recognition – fundamentals and applications*. John Wiley and Sons, 2001.

[25] A.E. Carpenter, T.R. Jones, M.R. Lamprecht, C. Clarke, I.H. Kang, O. Friman, D.A. Guertin, J.H. Chang, R.A. Lindquist, J. Moffat, P. Golland, and D.M. Sabatini. Cell profiler: image analysis software for identifying and quantifying cell phenotypes. *Genome Biology*, 7(R100), 2006.

[26] A.E. Carpenter and D.M. Sabatini. Systematic genome-wide screens of gene function. *Nature Reviews Genetics*, 5(1):11–22, 2004.

[27] C.-C. Chang and C.-J. Lin. *LIBSVM: a library for support vector machines*, 2001. Software available at http://www.csie.ntu.edu.tw/~cjlin/libsvm.

[28] T. Chang and C.C.J. Kuo. Texture analysis and classification with tree-structured wavelet transform. *IEEE Transactions on Image Processing*, 2(4):429–441, 1993.

[29] A. Chebira, J.A. Ozolek, C.A. Castro, W.G. Jenkinson, M. Gore, R. Bhagavatula, I. Khaimovich, S.E. Ormon, C.S. Navara, M. Sukhwani, K.E. Orwig, A. Ben-Yehudah, G. Schatten, G.K. Rohde, and J. Kovačević. Multiresolution identification of germ layer components in teratomas derived from human and nonhuman primate embryonic stem cells. In J.-C. Olivo-Marin, I. Bloch, and A. Laine, editors, *Proc. IEEE Internat. Symposium on Biomedical Imaging: From Nano to Macro (ISBI'2008)*, pages 979–982, Paris, France, May 14–17 2008.

[30] C. Chen, H. Li, X. Zhou, and S.T.C. Wong. Constraint factor graph cut-based active contour method for automated cellular image segmentation in RNAi screening. *Journal of Microscopy*, 230(2):177–191, 2008.

[31] P.-H. Chen, C.-J. Lin, and B. Schölkopf. A tutorial on ν-support vector machines: Research articles. *Applied Stochastic Models in Business and Industry*, 21(2):111–136, 2005.

[32] S.C. Chen, T. Zhao, G.J. Gordon, and R.F. Murphy. Automated image analysis of protein localization in budding yeast. *Bioinformatics*, 23(13):i66–i71, Jul 2007.

[33] X. Chen and R.F. Murphy. Robust classification of subcellular location patterns in high resolution 3D fluorescence microscope images. In *Proc. 26th Annu. Internat. Conf. IEEE Engineering in Medicine and Biology Society*, pages 1632–1635, San Francisco, CA, USA, 2004.

[34] X. Chen, M. Velliste, and R.F. Murphy. Automated interpretation of subcellular patterns in fluorescence microscope images for location proteomics. *Cytometry A*, 69A:631–640, 2006.

[35] X. Chen, X. Zhou, and S.T.C. Wong. Automated segmentation, classification, and tracking of cancer cell nuclei in time-lapse microscopy. *IEEE Transactions on Biomedical Engineering*, 53:762–766, 2006.

[36] Y. Chen, E. Ladi, P. Herzmark, E. Robey, and B. Roysam. Automated 5-D analysis of cell migration and interaction in the thymic cortex from time-lapse sequences of 3-D multi-channel multi-photon images. *Journal of Immunological Methods*, 340:65–80, 2009.

[37] Y. Cheng. Mean shift, mode seeking, and clustering. *IEEE Transactions on Pattern Analysis and Machine Intelligence*, 17(8):790–799, 1995.

[38] D. Chetverikov and J. Verestoy. Tracking feature points: a new algorithm. In *Proc. 14th Internat. Conf. Pattern Recognition*, volume 2, pages 1436–1438, Brisbane, Qld., Australia, Aug 1998.

[39] P. Comon. Independent component analysis - a new concept? *Signal Processing*, 36:287–314, 1994.

[40] C. Conrad. *High content screening on human cell arrays by automated microscopy and pattern recognition*. PhD thesis, Ruprecht-Karls-Universität Heidelberg, 2004.

[41] C. Conrad, H. Erfle, P. Warnat, N. Daigle, T. Lörch, J. Ellenberg, R. Pepperkok, and R. Eils. Automatic identification of subcellular phenotypes on human cell arrays. *Genome Research*, 14:1130–1136, 2004.

[42] C.F. Cullen, P. Deák, D.M. Glover, and H. Ohkura. mini spindles: A gene encoding a conserved microtubule-associated protein required for the integrity

of the mitotic spindle in drosophila. *Journal of Cell Biology*, 146:1005–1018, 1999.

[43] A. Danckaert, E. Gonzalez-Couto, L. Bollondi, N. Thompson, and B. Hayes. Automated recognition of intracellular organelles in confocal microscope images. *Traffic*, 3:66–73, 2002.

[44] M. Dash and H. Liu. Feature selection for classification. *Intelligent Data Analysis*, 1(3):131–156, 1997.

[45] M. Datar, D. Padfield, and H. Cline. Color and texture based segmentation of molecular pathology images using HSOMs. In J.-C. Olivo-Marin, I. Bloch, and A. Laine, editors, *Proc. IEEE Internat. Symposium on Biomedical Imaging: From Nano to Macro (ISBI'2008)*, pages 1343–1346, Paris, France, May 14–17 2008.

[46] I. Daubechies. Orthonormal bases of compactly supported wavelets. *Communications on Pure and Applied Mathematics*, 41:909–996, 1988.

[47] O. Debeir, P. Van Ham, R. Kiss, and C. Decaestecker. Tracking of migrating cells under phase-contrast video microscopy with combined mean-shift processes. *IEEE Transactions on Medical Imaging*, 24(6):697–711, 2005.

[48] E. Dimitriadou. *cclust: Convex Clustering Methods and Clustering Indexes*, 2002. R package version 0.6-9.

[49] E. Dimitriadou, K. Hornik, F. Leisch, D. Meyer, and A. Weingessel. *e1071: Misc Functions of the Department of Statistics*. TU Wien, 2005. R package version 1.5-8.

[50] A. Dufour, V. Shinin, S. Tajbakhsh, N. Guillén-Aghion, J.-C. Olivo-Marin, and C. Zimmer. Segmenting and tracking fluorescent cells in dynamic 3-D microscopy with coupled active surfaces. *IEEE Transactions on Image Processing*, 14(9):1396–1410, 2005.

[51] O. Dzyubachyk, W. Niessen, and E. Meijering. A variational model for level-set based cell tracking in time-lapse fluorescence microscopy images. In J. Fessler and T. Denney, editors, *Proc. IEEE Internat. Symposium on Biomedical Imaging: From Nano to Macro (ISBI'2007)*, pages 97–100, Arlington, VA, USA, Apr 12–15 2007.

[52] C.J. Echeverri and N. Perrimon. High-throughput RNAi screening in cultured cells: a user's guide. *Nature Reviews Genetics*, 7(5):373–384, 2006.

[53] H. Erfle, J.C. Simpson, P.I. Bastiaens, and R. Pepperkok. siRNA cell arrays for high-content screening microscopy. *Biotechniques*, 37(3):454–458, 460, 462, 2004.

[54] D. Fenistein, B. Lenseigne, T. Christophe, P. Brodin, and A. Genovesio. A fast, fully automated cell segmentation algorithm for high-throughput and high-content screeining. *Cytometry A*, 73:958–964, 2008.

[55] L. Ficsor, V.S. Varga, A. Tagscherer, Z. Tulassay, and B. Molnar. Automated classification of inflammation in colon histological sections based on digital microscopy and advanced image analysis. *Cytometry A*, (73):230–237, 2008.

[56] R. Fletcher. *Practical Methods of Optimization*. John Wiley and Sons, 2nd edition, 1987.

[57] G. Gallardo, F. Yang, F. Ianzini, M.A. Mackey, and M. Sonka. Mitotic cell recognition with hidden markov models. In Jr. R L Galloway, editor, *Medical Imaging 2004: Visualization, Image-Guided Procedures, and Display, Proc. SPIE*, volume 5367, pages 661–668, 2004.

[58] A.E. Gambe, R.M. Ono, S. Matsunaga, N. Kutsuna, T. Higaki, T. Higashi, S. Hasezawa, S. Uchiyama, and K. Fukui. Development of a multistage classifier for a monitoring system of cell activity based on imaging of chromosomal dynamics. *Cytometry A*, 71:286–296, 2007.

[59] D.L. Gard and M.W. Kirschner. A microtuble-associated protein from Xenopus eggs that specifically promotes assembly at the plus-end. *Journal of Cell Biology*, 105:2203–2215, 1987.

[60] M. Gipp, G. Marcus, N. Harder, A. Suratanee, K. Rohr, R. König, and R. Männer. Accelerating the computation of Haralick's texture features using graphics processing units (GPUs). In *Proc. World Congress on Engineering 2008 (WCE'08), The 2008 Internat. Conf. of Parallel and Distributed Computing (ICPDC'08)*, pages 587–593, London, UK, Jul 2–4 2008. Newswood Limited, International Association of Engineers.

[61] M. Gipp, G. Marcus, N. Harder, A. Suratanee, K. Rohr, R. König, and R. Männer. Haralick's texture features computed by GPUs for biological applications. *IAENG International Journal of Computer Science*, 36(1), 2009.

[62] D. Glotsos, I. Kalatzis, P. Spyridonos, S. Kostopoulos, A. Daskalakis, E. Athanasiadis, P. Ravazoula, G. Nikiforidis, and D. Cavouras. Improving

accuracy in astrocytomas grading by integrating a robust least squares mapping driven support vector machine classifier into a two level grade classification scheme. *Computer Methods and Programs in Biomedicine*, 90(3):251–261, 2008.

[63] W.J. Godinez. Tracking mitotic cells in 2D and 3D cell microscopy images, Aug 2005. Bachelor's thesis, International University in Germany, Bruchsal.

[64] R.C. Gonzalez and R.E. Woods. *Digital Image Processing*. Prentice Hall, 2nd edition, Jan 2002.

[65] A. Gordon, A. Colman-Lerner, T.E. Chin, K.R. Benjamin, R.C. Yu, and R. Brent. Single-cell quantification of molecules and rates using open-source microscope-based cytometry. *Nature Methods*, 4(2):175–181, Jan 2007.

[66] N.J. Gordon, D.J. Salmond, and A.F.M. Smith. Novel approach to nonlinear/non-gaussian bayesian state estimation. *Radar and Signal Processing, IEE Proceedings F*, 140(2):107–113, 1993.

[67] G. Goshima, R. Wollman, S.S. Goodwin, N. Zhang, J.M. Scholey, R.D. Vale, and N. Stuurman. Genes required for mitotic spindle assembly in drosophila S2 cells. *Science*, 316(5823):417–421, Apr 2007.

[68] N. Grabe, T. Pommerencke, T. Steinberg, H. Dickhaus, and P. Tomakidi. Reconstructing protein networks of epithelial differentiation from histological sections. *Bioinformatics*, 23(23):3200–3208, Dec 2007.

[69] N.A. Hamilton, R.S. Pantelic, K. Hanson, and R.D. Teasdale. Fast automated cell phenotype image classification. *BMC Bioinformatics*, 8(110), Mar 2007.

[70] J. Han, H. Chang, K. Andarawewa, P. Yaswen, M.H. Barcellos-Hoff, and B. Parvin. Integrated profiling of cell surface protein and nuclear marker for discriminant analysis. In J.-C. Olivo-Marin, I. Bloch, and A. Laine, editors, *Proc. IEEE Internat. Symposium on Biomedical Imaging: From Nano to Macro (ISBI'2008)*, pages 1343–1346, Paris, France, May 14–17 2008.

[71] J.N. Harada, K.E. Bower, A.P. Orth, S. Callaway, C.G. Nelson, C. Laris, J.B. Hogenesch, P.K. Vogt, and S.K. Chanda. Identification of novel mammalian growth regulatory factors by genome-scale quantitative image analysis. *Genome Research*, 15(8):1136–1144, Jul 2005.

[72] R. M. Haralick. Statistical and structural approaches to texture. *Proceedings of the IEEE*, 67(5):768–804, 1979.

[73] N. Harder, R. Eils, and K. Rohr. Automated classification of mitotic phenotypes of human cells using fluorescent proteins. In K.F. Sullivan, editor, *Fluorescent Proteins*, volume 85 of *Methods in Cell Biology*, pages 539–554. Academic Press, 2008.

[74] N. Harder, F. Mora-Bermúdez, W.J. Godinez, J. Ellenberg, R. Eils, and K. Rohr. Automated analysis of the mitotic phases of human cells in 3D fluorescence microscopy image sequences. In R. Larsen, M. Nielsen, and J. Sporring, editors, *Proc. 9th Internat. Conf. on Medical Image Computing and Computer-Assisted Intervention (MICCAI'2006)*, volume 4190 of *LNCS*, pages 840–848, Copenhagen, DK, Oct 1–6 2006. Springer-Verlag.

[75] N. Harder, F. Mora-Bermúdez, W.J. Godinez, J. Ellenberg, R. Eils, and K. Rohr. Automated analysis of the mitotic phases of human cells in 3D fluorescence microscopy image sequences. In D.N. Metaxas, R.T. Whitaker, J. Rittscher, and T.B. Sebastian, editors, *Proc. MICCAI'06 Workshop Microscopic Image Analysis with Applications in Biology (MIAAB'2006)*, Copenhagen, DK, Oct 5 2006.

[76] N. Harder, F. Mora-Bermúdez, W.J. Godinez, J. Ellenberg, R. Eils, and K. Rohr. Determination of mitotic delays in 3D fluorescence microscopy images of human cells using an error-correcting finite state machine. In A. Horsch, T.M. Deserno, H. Handels, H.-P. Meinzer, and T.Tolxdorff, editors, *Proc. Workshop Bildverarbeitung für die Medizin 2007 (BVM'07)*, Informatik aktuell, pages 242–246, Munich, Germany, Mar 25–27 2007. Springer-Verlag.

[77] N. Harder, F. Mora-Bermúdez, W.J. Godinez, J. Ellenberg, R. Eils, and K. Rohr. Determination of mitotic delays in 3D fluorescence microscopy images of human cells using an error-correcting finite state machine. In J. Fessler and T. Denney, editors, *Proc. IEEE Internat. Symposium on Biomedical Imaging: From Nano to Macro (ISBI'2007)*, pages 1044–1047, Arlington, VA, USA, Apr 12–15 2007.

[78] N. Harder, F. Mora-Bermúdez, W.J. Godinez, J. Ellenberg, R. Eils, and K. Rohr. Automated analysis of mitotic cell nuclei in 3D fluorescence microscopy image sequences. In B.S. Manjunath, G. Danuser, and A. Carpenter, editors, *Workshop on Bio-Image Informatics: Biological Imaging, Computer Vision and Data Mining*, Center for Bio-Image Informatics, UCSB, Santa Barbara, CA, USA, Jan 17–18 2008.

[79] N. Harder, F. Mora-Bermúdez, W.J. Godinez, J. Ellenberg, R. Eils, and K. Rohr. Automated analysis of the mitotic phases of human cells in 3D fluorescence microscopy image sequences. In J. Rittscher, R. Machiraju, and S.T.C. Wong, editors, *Microscopic Image Analysis for Life Science Applications*, pages 283–293. Artech House, 2008.

[80] N. Harder, F. Mora-Bermúdez, W.J. Godinez, A. Wünsche, R. Eils, J. Ellenberg, and K. Rohr. Automatic analysis of dividing cells in live cell movies to detect mitotic delays and correlate phenotypes in time. *Genome Research*, 19(11):2113–2124, Nov 2009.

[81] N. Harder, B. Neumann, M. Held, U. Liebel, H. Erfle, J. Ellenberg, R. Eils, and K. Rohr. Automated recognition of mitotic patterns in fluorescence microscopy images of human cells. In J. Kovačević and E. Meijering, editors, *Proc. IEEE Internat. Symposium on Biomedical Imaging: From Nano to Macro (ISBI'2006)*, pages 1016–1019, Arlington, VA, USA, Apr 6–9 2006.

[82] N. Harder, B. Neumann, M. Held, U. Liebel, H. Erfle, J. Ellenberg, R. Eils, and K. Rohr. Automated recognition of mitotic phenotypes in fluorescence microscopy images of human cells. In H. Handels, J. Ehrhardt, A. Horsch, H.-P. Meinzer, and T. Tolxdorff, editors, *Proc. Workshop Bildverarbeitung für die Medizin 2006 (BVM'06)*, Informatik aktuell, pages 206–210, Hamburg, Germany, Mar 19–21 2006. Springer-Verlag.

[83] D.D. Hearn and M.P. Baker. *Computer Graphics with OpenGL*. Prentice Hall, 3rd edition, Sep 2003.

[84] R.M. Heiberger and B. Holland. *Statistical Analysis and Data Display*. Springer Texts in Statistics. Springer-Verlag, 2004.

[85] P. Hong, T.S. Huang, and M. Turk. Gesture modeling and recognition using finite state machines. In *Proc. IEEE Internat. Conference on Automatic Face and Gesture Recognition (FG 2000)*, pages 410–415, Grenoble, France, Mar 26–30 2000. IEEE Computer Society.

[86] J.E. Hopcroft and J.D. Ullman. *Introduction to Automata Theory, Languages, and Computation*. Addison-Wesley, Reading, MA, USA, 1st edition, 1979.

[87] C.-W. Hsu, C.-C. Chang, and C.-J. Lin. *A Practical Guide to Support Vector Classification*, 2001. http://www.csie.ntu.edu.tw/~cjlin.

[88] C.-W. Hsu and C.-J. Lin. A comparison of methods for multiclass support vector machines. *IEEE Transactions on Neural Networks*, 13(2):415–425, 2002.

[89] K. Huang and R.F. Murphy. Boosting accuracy of automated classification of fluorescence microscope images for location proteomics. *BMC Bioinformatics*, 5, Jun 2004.

[90] K. Huang and R.F. Murphy. From quantitative microscopy to automated image understanding. *Journal of Biomedical Optics*, 9(5):893–912, Sep–Oct 2004.

[91] K. Huang, M. Velliste, and R.F. Murphy. Feature reduction for improved recognition of subcellular location patterns in fluorescence microscopy images. In D.V. Nicolau, J. Enderlein, R.C. Leif, and D.L. Farkas, editors, *Medical Imaging 2003: Manipulation and Analysis of Biomolecules, Cells, and Tissues, Proc. SPIE*, volume 4962, pages 307–318, 2003.

[92] A. Hyvärinen. Fast and robust fixed-point algorithms for independent component analysis. *IEEE Transactions of Neural Networks*, 10(3):626–634, 1999.

[93] I.T. Jolliffe. *Principal Component Analysis*. Springer-Verlag, New York, 1986.

[94] T.R. Jones, A.E. Carpenter, and P. Golland. Voronoi-based segmentation of cells on image manifolds. In Y. Liu, T. Jiang, and C. Zhang, editors, *Computer Vision for Biomedical Image Applications, First Internat. Workshop (CVBIA'05)*, volume 3765 of *LNCS*, pages 535–543, Beijing, China, Oct 21 2005. Springer-Verlag.

[95] T.R. Jones, A.E. Carpenter, D.M. Sabatini, and P. Golland. Methods for high-content, high-throughput image-based cell screening. In D.N. Metaxas, R.T. Whitaker, J. Rittscher, and T.B. Sebastian, editors, *Proc. MICCAI'06 Workshop Microscopic Image Analysis with Applications in Biology (MIAAB'2006)*, Copenhagen, DK, Oct 5 2006.

[96] R. E. Kalman. A new approach to linear filtering and prediction problems. *Transactions of the ASME – Journal of Basic Engineering*, (82 (Series D)):35–45, 1960.

[97] M. Kass, A. Witkin, and D. Terzopoulos. Snakes: Active contour models. *International Journal of Computer Vision*, pages 321–331, 1988.

[98] A. Khotanzad and Y.H. Hong. Rotation invariant image recognition using features selected via a systematic method. *Pattern Recognition*, 23(10):1089–1101, 1990.

[99] D. Kim, Y.-S. Chae, and S.J. Kim. High content cellular analysis for functional screening of novel cell cycle related genes. In *Proc. International Conference on BioMedical Engineering and Informatics (BMEI'2008)*, pages 148–152, May 27–30 2008.

[100] I.-H. Kim, W.J. Godinez, N. Harder, F. Mora-Bermúdez, J. Ellenberg, R. Eils, and K. Rohr. Compensation of global movement for improved tracking of cells in time-lapse confocal microscopy image sequences. In J.P.W. Pluim and J.M. Reinhardt, editors, *Medical Imaging 2007: Image Processing (MI'07)*, *Proc. SPIE*, volume 6512, San Diego, CA, USA, Feb 17–22 2007.

[101] J. Kittler and J. Illingworth. Minimum error thresholding. *Pattern Recognition*, 19(1):41–47, 1986.

[102] S. Knerr, L. Personnaz, and G. Dreyfus. Single-layer learning revisted: A stepwise procedure for building and training a neural network. In F. Fogelman and J. Hérault, editors, *Neurocomputing: Algorithms, Architectures and Applications*. Springer-Verlag, 1990.

[103] V. Kovalev, N. Harder, B. Neumann, M. Held, U. Liebel, H. Erfle, J. Ellenberg, R. Eils, and K. Rohr. Feature selection for evaluating fluorescence microscopy images in genome-wide cell screens. In C. Schmid, S. Soatto, and C. Tomasi, editors, *Proc. IEEE Computer Society Conf. on Computer Vision and Pattern Recognition (CVPR'06)*, pages 276–283, New York, NY, USA, Jun 17–22 2006.

[104] U. Kreßel. Pairwise classification and support vector machines. In B. Schölkopf, C.J.C. Burges, and A.J. Smola, editors, *Advances in Kernel Methods–Support Vector Learning*, pages 255–268. The MIT press, 1999.

[105] M.R. Lamprecht, D.M. Sabatini, and A.E. Carpenter. Cellprofiler: free, versatile software for automated biological image analysis. *Biotechniques*, 42(1):71–75, 2007.

[106] G.H. Landeweerd and E.S. Gelsema. The use of nuclear texture parameters in the automatic analysis of leukocytes. *Pattern Recognition*, 10(2):57–61, 1978.

[107] H.B. Landsverk, M. Kirkhus, M. Bollen, T. Küntziger, and P. Collas. PNUTS enhances in vitro chromosome decondensation in a PP1-dependent manner. *Biochemical Journal*, 390:709–717, 2005.

[108] P. Leray and P. Gallinari. Feature selection with neural networks. *Behaviormetrika*, 26(1):145–166, 1999.

[109] F. Li, X. Zhou, J. Ma, and S.T.C. Wong. An automated feedback system with hybrid model of scoring and classification for solving over-segmentation problems in RNAi high content screening. *Journal of Microscopy*, 226:121–132, Jan 2007.

[110] K. Li and T. Kanade. Cell population tracking and lineage construction using multiple-model dynamics filters and spatiotemporal optimization. In D.N. Metaxas, J. Rittscher, S. Lockett, and T.B. Sebastian, editors, *Proc. MICCAI'07 Workshop Microscopic Image Analysis with Applications in Biology (MIAAB'2007)*, Piscataway, NJ, USA, 2007.

[111] K. Li, E.D. Miller, M. Chen, T. Kanade, L.E. Weiss, and P.G. Campbell. Cell population tracking and lineage construction with spatiotemporal context. *Medical Image Analysis*, 12(5):546–566, 2008.

[112] K. Li, E.D. Miller, M. Chen, T. Kanade, L.E. Weiss, and P.G. Campbell. Computer vision tracking of stemness. In J.-C. Olivo-Marin, I. Bloch, and A. Laine, editors, *Proc. IEEE Internat. Symposium on Biomedical Imaging: From Nano to Macro (ISBI'2008)*, pages 847–850, Paris, France, May 14–17 2008.

[113] K. Li, E.D. Miller, L.E. Weiss, P.G. Campbell, and T. Kanade. Online tracking of migrating and proliferating cells imaged with phase-contrast microscopy. In *Proc. IEEE Workshop on Mathematical Methods in Biomedical Image Analysis (MMBIA'2006)*, pages 65–72, New York, NY, USA, Jun 2006.

[114] A. Liaw and M. Wiener. Classification and regression by randomForest. *R News*, 2(3):18–22, 2002.

[115] G. Lin, U. Adiga, K. Olson, J. Guzowski, C. Barnes, and B. Roysam. A hybrid 3D watershed algorithm incorporating gradient cues and object models for automatic segmentation of nuclei in confocal image stacks. *Cytometry A*, 56:23–36, 2003.

[116] G. Lin, M.K. Chawla, K. Olson, C.A. Barnes, J.F. Guzowski, C. Bjornsson, W. Shain, and B. Roysam. A multi-model approach to simultaneous segmentation and classification of heterogeneous populations of cell nuclei in 3D confocal microscope images. *Cytometry A*, 71:724–736, 2007.

[117] J. Lindblad, C. Wählby, E. Bengtsson, and A. Zaltsman. Image analysis for automatic segmentation of cytoplasms and classification of Rac1 activation. *Cytometry A*, 57A:22–33, 2003.

[118] Y. Liron, Y. Paran, I. Lavelin, S. Naffar-Abu-Amara, S. Winograd-Katz, B. Geiger, and Z. Kam. Image aquisition and understanding in high-throughput high-resolution cell-based screening applications. In J.-C. Olivo-Marin, I. Bloch, and A. Laine, editors, *Proc. IEEE Internat. Symposium on Biomedical Imaging: From Nano to Macro (ISBI'2008)*, pages 332–335, Paris, France, May 14–17 2008.

[119] L.-H. Loo, L.F. Wu, and S.J. Altschuler. Image-based multivariate profiling of drug responses from single cells. *Nature Methods*, 4(5):445–453, May 2007.

[120] S. Mallat. *A wavelet tour of signal processing*. Academic Press, 1998.

[121] N. Malpica, C. Ortiz de Solórzano, J.J. Vaquero, A. Santos, I. Vallcorba, J.M. García-Sagredo, and F. del Pozo. Applying watershed algorithms to the segmentation of clustered nuclei. *Cytometry*, 28:289–297, 1997.

[122] H.B. Mann and D.R. Whitney. On a test of whether one of two random variables is stochastically larger than the other. *Annals of Mathematical Statistics*, 18(1):50–60, 1947.

[123] P. Matula, A. Kumar, I. Wörz, H. Erfle, R. Bartenschlager, R. Eils, and K. Rohr. Single-cell-based image analysis of high-throughput cell array screens for quantification of viral infection. *Cytometry A*, 75:309–318, 2009.

[124] P. Matula, A. Kumar, I. Wörz, N. Harder, H. Erfle, R. Bartenschlager, R. Eils, and K. Rohr. Automated analysis of siRNA screens of cells infected by Hepatitis C and Dengue viruses based on immunofluorescence microscopy images. In X.P. Hu and A.V. Clough, editors, *Medical Imaging 2008: Physiology, Function, and Structure from Medical Images (MI'08)*, Proc. SPIE, San Diego, CA, USA, Feb 16–21 2008.

[125] P. Matula, A. Kumar, I. Wörz, N. Harder, H. Erfle, R. Bartenschlager, R. Eils, and K. Rohr. Automated analysis of siRNA screens of virus infected cells based

on immunofluorescence microscopy images. In T. Tolxdorff, J. Braun, T.M. Deserno, H. Handels, A. Horsch, and H.-P. Meinzer, editors, *Proc. Workshop Bildverarbeitung für die Medizin 2008 (BVM'08)*, Informatik aktuell, pages 453–457, Berlin, Germany, Apr 6–8 2008. Springer-Verlag.

[126] T. McInerney and D. Terzopoulos. Deformable models in medical image analysis: A survey. *Medical Image Analysis*, 1:91–108, 1996.

[127] M. Minsky. *Computation: Finite and Infinite Machines.* Prentice Hall, New Jersey, NY, USA, 1st edition, 1967.

[128] M. Mohri. Finite-state transducers in language and speech processing. *Computational Linguistics*, 23(2):269–311, 1997.

[129] F. Mora-Bermúdez. *Quantitative Analysis of the Structural Dynamics of Mitotic Chromosomes in Live Mammalian Cells.* PhD thesis, Ruprecht-Karls-Universität Heidelberg, 2006.

[130] F. Mora-Bermúdez and J. Ellenberg. Measuring structural dynamics of chromosomes in living cells by fluorescence microscopy. *Methods*, 41:158–167, 2007.

[131] D.P. Mukherjee, N. Ray, and S. Acton. Level set analysis of leukocyte detection and tracking. *IEEE Transactions on Image Processing*, 13(4):562–572, 2004.

[132] R.F. Murphy, M. Velliste, and G. Porreca. Robust classification of subcellular location patterns in fluorescence microscope images. In *Proc. 2002 IEEE Internat. Workshop on Neural Networks for Signal Processing (NNSP 12)*, pages 67–76, Sep 4–6 2002.

[133] S.K. Nath, K. Palaniappan, and F. Bunyak. Cell segmentation using coupled level sets and graph-vertex coloring. In R. Larsen, M. Nielsen, and J. Sporring, editors, *Proc. 9th Internat. Conf. on Medical Image Computing and Computer-Assisted Intervention (MICCAI'2006)*, volume 4190 of *LNCS*, pages 101–108, Copenhagen, DK, Oct 1–6 2006. Springer-Verlag.

[134] T.W. Nattkemper, H.J. Ritter, and W. Schubert. A neural classifier enabling high-throughput topological analysis of lymphocytes in tissue sections. *IEEE Transactions on Information Technology in Biomedicine*, 5(2):138–149, 2001.

[135] T.W. Nattkemper, W. Schubert, T. Hermann, and H.J. Ritter. A hybrid system for cell detection in digital micrographs. In B. Tilg, editor, *Biomedical Engineering, Proc. BIOMED 2004*, volume 417, Innsbruck, Austria, Feb 2004. ACTA Press.

[136] B. Neumann, M. Held, U. Liebel, H. Erfle, P. Rogers, R. Pepperkok, and J. Ellenberg. High-throughput RNAi screening by time-lapse imaging of live human cells. *Nature Methods*, 3(5):385–390, May 2006.

[137] C. Ortiz de Solórzano, R. Malladi, S.A. Lelièvre, and S.J. Lockett. Segmentation of nuclei and cells using membrane related protein markers. *Journal of Microscopy*, 201(3):404–415, 2001.

[138] S. Osher and J.A. Sethian. Fronts propagating with curvature-dependent speed: algorithmns based on Hamilton-Jacobi formulations. *Journal of Computational Physics*, 79:12–49, 1988.

[139] E.E. Osuna, R. Freund, and F. Girosi. Support vector machines: Training and applications. A.I. Memo 1602, Massachusetts Institute of Technology, 1997.

[140] N. Otsu. A threshold selection method from grey level histograms. *IEEE Transactions on Systems, Man and Cybernetics*, 9:62–66, 1979.

[141] D.R. Padfield, J. Rittscher, N.Thomas, and B. Roysam. Spatio-temporal cell cycle phase analysis using level sets and fast marching methods. *Medical Image Analysis*, 13:143–155, 2009.

[142] D.R. Padfield, J. Rittscher, and B. Roysam. Spatio-temporal cell segmentation and tracking for automated screening. In J.-C. Olivo-Marin, I. Bloch, and A. Laine, editors, *Proc. IEEE Internat. Symposium on Biomedical Imaging: From Nano to Macro (ISBI'2008)*, pages 376–379, Paris, France, May 14–17 2008.

[143] D.R. Padfield, J. Rittscher, T. Sebastian, N. Thomas, and B. Roysam. Spatio-temporal cell cycle analysis using 3D level set segmentation of unstained nuclei in line scan confocal fluorescence images. In J. Kovačević and E. Meijering, editors, *Proc. IEEE Internat. Symposium on Biomedical Imaging: From Nano to Macro (ISBI'2006)*, pages 1036–1039, Arlington, VA, USA, Apr 6–9 2006.

[144] D.R. Padfield, J. Rittscher, N. Thomas, and B. Roysam. Spatio-temporal cell cycle phase analysis using level sets and fast marching methods. In D.N. Metaxas, R.T. Whitaker, J. Rittscher, and T.B. Sebastian, editors, *Proc. MICCAI'06 Workshop Microscopic Image Analysis with Applications in Biology (MIAAB'2006)*, pages 2–9, 2006.

[145] Z.E. Perlman, T.J. Mitchison, and T.U. Mayer. High-content screening and profiling of drug activity in an automated centrosome-duplication assay. *Chem-BioChem*, 6(1):145–151, 2005.

[146] Z.E. Perlman, M.D. Slack, Y. Feng, T.J. Mitchison, L.F. Wu, and S.J. Altschuler. Multidimensional drug profiling by automated microscopy. *Science*, 306:1194–1198, 2004.

[147] D. Perrin. Finite automata. In J. van Leeuwen, editor, *Handbook of Theoretical Computer Science, Volume B: Formal Models and Semantics*. Elsevier and MIT Press, 1990.

[148] T.D. Pham and D.T. Tran. Gaussian mixture and markov models for cell-phase classification in microscopic images. In *Proc. IEEE/SMC Internat. Conference on System of Systems Engineering*, pages 328–333, Los Angeles, CA, USA, Apr 2006.

[149] T.D. Pham, D.T. Tran, X. Zhou, and S.T.C. Wong. Integrated algorithms for image analysis and classification of nuclear division for high-content cell-cycle screening. *International Journal of Computational Intelligence and Applications*, 6(1):21–43, 2006.

[150] Z. Pincus and J.A. Theriot. Comparison of quantitative methods for cell-shape analysis. *Journal of Microscopy*, 227:140–156, Mar 2007.

[151] J.C. Platt, N. Cristianini, and J. Shawe-Taylor. Large margin DAG's for multiclass classification. In *Advances in Neural Information Processing Systems*, volume 12, pages 547–553. The MIT press, 2000.

[152] R.J. Prokop and A.P. Reeves. A survey of moment-based techniques for unoccluded object representation and recognition. *CVGIP: Graphical Models and Image Processing*, 54(5):438–460, 1992.

[153] D. Puig and M.A. Garcia. Automatic texture feature selection for image pixel classification. *Pattern Recognition*, 39:1996–2009, 2006.

[154] R Development Core Team. *R: A Language and Environment for Statistical Computing*. R Foundation for Statistical Computing, Vienna, Austria, 2008.

[155] V. Racine, A. Hertzog, J. Jouanneau, J. Salamero, C. Kervrann, and J.-B. Sibarita. Multiple-target tracking of 3d fluorescent objects based on simulated annealing. In J. Kovačević and E. Meijering, editors, *Proc. IEEE Internat.*

Symposium on Biomedical Imaging: From Nano to Macro (ISBI'2006), pages 1020–1023, Arlington, VA, USA, Apr 6–9 2006.

[156] W.S. Rasband. *ImageJ*. National Institutes of Health, Bethesda, Maryland, USA, 1997–2004. Software available at http://rsb.info.nih.gov/ij/.

[157] E. Roche and Y. Shabes, editors. *Finite-State Language Processing*. MIT Press, Cambridge, MA, USA, 1997.

[158] K. Rodenacker and E. Bengtsson. A feature set for cytometry on digitized microscopic images. *Analytical Cellular Pathology*, 25(1):1–36, 2003.

[159] O. Ronneberger, E. Schultz, and H. Burkhardt. Automated pollen recognition using 3D volume images from fluorescence microscopy. *Aerobiologia*, 18:107–115, 2002.

[160] M. Roula, A. Bouridan, F. Kurugollu, and J. Diamond. 3D segmentation and feature extraction of CLSM scanned nuclei using evolutionary snakes. In J. Fessler and T. Denney, editors, *Proc. IEEE Internat. Symposium on Biomedical Imaging: From Nano to Macro (ISBI'2007)*, pages 316–319, Arlington, VA, USA, Apr 12–15 2007.

[161] J. Sethian. *Level Set Methods and Fast Marching Methods*. Cambridge University Press, Cambridge, 2nd edition, 1999.

[162] M. Sezgin and B. Sankur. Survey over image thresholding techniques and quantitative performance evaluation. *Journal of Electronic Imaging*, 13(1):146–168, Jan 2004.

[163] C.E. Shannon. A mathematical theory of communication. *The Bell System Technical Journal*, 27:379–423 and 623–656, Jul and Oct 1948.

[164] S.S. Shapiro and M.B. Wilk. An analysis of variance test for normality (complete samples). *Biometrika*, 52(3 and 4):591–611, 1965.

[165] A. Sigal, R. Milo, A. Cohen, N. Geva-Zatorsky, Y. Klein, I. Alaluf, N. Swerdlin, N. Perzov, T. Danon, Y. Liron, T. Raveh, A.E. Carpenter, G. Lahav, and U. Alon. Dynamic proteomics in individual human cells uncovers widespread cell-cycle dependence of nuclear proteins. *Nature Methods*, 3(4):525–531, Jul 2006.

[166] N. Stuurman. ImageJ plugin MTrack2, 2003. Software available at http://valelab.ucsf.edu/~nico/IJplugins/MTrack2.html.

[167] A.T. Szafran, M. Marcelli, and M.A. Mancini. High throughput multiplex image analyses for androgen receptor function. In J.-C. Olivo-Marin, I. Bloch, and A. Laine, editors, *Proc. IEEE Internat. Symposium on Biomedical Imaging: From Nano to Macro (ISBI'2008)*, pages 320–323, Paris, France, May 14–17 2008.

[168] C.Y. Tao, J. Hoyt, and Y. Feng. A support vector machine classifier for recognizing mitotic subphases using high-content screening data. *Journal of Biomolecular Screening*, 12(4):490–496, 2007.

[169] S. Theodoridis and K. Koutroumbas. *Pattern Recognition*. Academic Press, 1999.

[170] T.M. Therneau and B. Atkinson. *R port by B. Ripley. rpart: Recursive Partitioning*, 2005. R package version 3.1-23.

[171] V.G. Tusher, R. Tibshirani, and G. Chu. Significance analysis of microarrays applied to the ionizing radiation response. *PNAS*, 98(9):5116–5121, 2001.

[172] V. Vapnik. *The Nature of Statistical Learning Theory*. Springer-Verlag, New York, 1995.

[173] V. Vapnik. *Statistical Learning Theory*. John Wiley and Sons, New York, 1998.

[174] M. Velliste and R.F. Murphy. Automated determination of protein subcellular locations from 3D fluorescence microscopy images. In M. Unser and Z.-P. Liang, editors, *Proc. IEEE Internat. Symposium on Biomedical Imaging: From Nano to Macro (ISBI'2002)*, pages 867–870, Washington, DC, USA, Jul 7–10 2002.

[175] L. Vincent and P. Soille. Watersheds in digital spaces: an efficient algorithm based on immersion simulations. *IEEE Transactions on Pattern Analysis and Machine Intelligence*, 13:583–598, 1991.

[176] C. Wählby, I.-M. Sintorn, F. Erlandsson, G. Borgefors, and E. Bengtsson. Combining intensity, edge and shape information for 2D and 3D segmentation of cell nuclei in tissue sections. *Journal of Microscopy*, 215:67–76, Jan 2004.

[177] T. Walter, M. Held, B. Neumann, J.-K. Hériché, C. Conrad, R. Pepperkok, and J. Ellenberg. A genome wide RNAi screen by time lapse microscopy in order to

identify mitotic genes – computational aspects and challenges. In J.-C. Olivo-Marin, I. Bloch, and A. Laine, editors, *Proc. IEEE Internat. Symposium on Biomedical Imaging: From Nano to Macro (ISBI'2008)*, pages 328–331, Paris, France, May 14–17 2008.

[178] J. Wang, X. Zhou, P.L. Bradley, S.-F. Chang, N. Perrimon, and S.T.C. Wong. Cellular phenotype recognition for high-content RNA interference genome-wide screening. *Journal of Biomolecular Screening*, 13(1):29–39, 2008.

[179] M. Wang, J. Yang, G.-P. Liu, Z.-J. Xu, and K.-C. Chou. Weighted-support vector machines for predicting membrane protein types based on pseudo-amino acid composition. *Protein Engineering, Design & Selection*, 17:509–516, 2004.

[180] M. Wang, X. Zhou, R.W. King, and S.T.C. Wong. Context based mixture model for cell phase identification in automated fluorescence microscopy. *BMC Bioinformatics*, 8(32), Jan 2007.

[181] M. Wang, X. Zhou, F. Li, J. Huckins, R.W. King, and S.T.C. Wong. Novel cell segmentation and online SVM for cell cycle phase identification in automated microscopy. *Bioinformatics*, 24(1):94–101, Jan 2008.

[182] X. Wang, W. He, D. Metaxas, R. Mathew, and E. White. Cell segmentation and tracking using texture adaptive snakes. In J. Fessler and T. Denney, editors, *Proc. IEEE Internat. Symposium on Biomedical Imaging: From Nano to Macro (ISBI'2007)*, pages 101–104, Arlington, VA, USA, Apr 12–15 2007.

[183] J. Weston and C. Watkins. Support vector machines for multi-class pattern recognition. In M. Verleysen, editor, *Proc. European Symposium on Artificial Neural Networks (ESANN'99)*, pages 219–224, 1999.

[184] D.B. Wheeler, S.N. Bailey, D.A. Guertin, A.E. Carpenter, C.O. Higgins, and D.M. Sabatini. RNAi living-cell microarrays for loss-of-function screens in Drosophila melanogaster cells. *Nature Methods*, 1(2):127–132, Oct 2004.

[185] J.A. Withers and K.A. Robbins. Tracking cell splits and merges. In *Proc. IEEE Southwest Symposium on Image Analysis and Interpretation*, pages 117–122, San Antonio, TX, USA, Apr 1996.

[186] W.A. Woods. Transition network grammars for natural language analysis. *Communications of the ACM*, 13(10):591–606, 1970.

[187] H.S. Wu, J. Gil, and L. Deligdisch. Region growing segmentation of chromatin clumps of ovarian cells using adaptive gradients. *Journal of Imaging Science and Technology*, 48:22–27, 2004.

[188] A. Wünsche. Comprehensive RNAi screening of human nuclear envelope proteins by newly developed high-throughput confocal time-lapse microscopy in living cells. Master's thesis, Ludwig-Maximilians-Universität München, 2007.

[189] T. Würflinger, J. Stockhausen, D. Meyer-Ebrecht, and A. Boecking. Robust automatic coregistration, segmentation, and classification of cell nuclei in multimodal cytopathological microscopic images. *Computerized Medical Imaging and Graphics*, 28:87–98, 2004.

[190] J. Yan, X. Zhou, Q. Yang, N. Liu, Q. Cheng, and S.T.C. Wong. An effective system for optical microscopy cell image segmentation, tracking and cell phase identification. In *Proc. IEEE Internat. Conference on Image Processing (ICIP'2006)*, pages 1917–1920, Atlanta, Georgia, USA, Oct 8–11 2006.

[191] F. Yang, M.A. Mackey, F. Ianzini, G. Gallardo, and M. Sonka. Cell segmentation, tracking, and mitosis detection using temporal context. In J. Duncan and G. Gering, editors, *Proc. 8th Internat. Conf. on Medical Image Computing and Computer-Assisted Intervention (MICCAI'2005)*, volume 3749 of *Lecture Notes in Computer Science*, pages 302–309, Palm Springs, CA, USA, Oct 2005. Springer-Verlag.

[192] X. Yang, H. Li, X. Zhou, and S.T.C. Wong. Identification of cell-cycle phases using neural network and steerable filter features. In *Proc. Third Internat. Symposium on Neural Networks (ISNN'2006), Advances in Neural Networks*, volume 3973 of *Lecture Notes in Computer Science*, pages 702–709. Springer-Verlag, 2006.

[193] M. Yeasin and S. Chaudhuri. Visual understanding of dynamic hand gestures. *Pattern Recognition*, 33:1805–1817, 2000.

[194] T.S. Yoo, M.J. Ackerman, W.E. Lorensen, W. Schroeder, V. Chalana, S. Aylward, D. Metaxes, and R. Whitaker. Engineering and algorithm design for an image processing API: A technical report on ITK - the insight toolkit. In J. Westwood, editor, *Proc. of Medicine Meets Virtual Reality*, pages 586–592. IOS Press Amsterdam, 2002. Website: "The Insight Segmentation and Registration Toolkit", http://www.itk.org.

[195] F. Zernike. Beugungstheorie des Schneidenverfahrens und seiner verbesserten Form, der Phasenkontrastmethode. *Physika*, 1:689–704, 1934.

[196] Y. Zhai, Z. Rasheed, and M. Shah. Conversation detection in feature films using finite state machines. In *Proc. IEEE Internat. Conference on Pattern Recognition (ICPR'2004)*, volume 4, pages 458–461, Cambridge, UK, Aug 23–26 2004.

[197] Y. Zhai, Z. Rasheed, and M. Shah. Semantic classification of movie scenes using finite state machines. In *IEE Proceedings – Vision, Image and Signal Processing*, volume 152, pages 896–901, Dec 2005.

[198] B. Zhang, C. Zimmer, and J.-C. Olivo-Marin. Tracking fluorescent cells with coupled geometric active contours. In R.M. Leahy and C. Roux, editors, *Proc. IEEE Internat. Symposium on Biomedical Imaging: From Nano to Macro (ISBI'2004)*, pages 476–479, Arlington, VA, USA, Apr 15–18 2004.

[199] X. Zhou, X. Cao, Z. Perlman, and S.T.C. Wong. A computerized cellular imaging system for high content analysis in monastrol suppressor screens. *Journal of Biomedical Informatics*, 39(2):115–125, Apr 2006.

[200] X. Zhou, F. Li, J. Yan, and S.T.C. Wong. A novel cell segmentation method and cell phase identification using Markov model. *IEEE Transactions on Information Technology in Biomedicine*, 13(2):152–157, 2009.

[201] X. Zhou, K.-Y. Liu, P. Bradley, N. Perrimon, and Stephen T.C. Wong. Towards automated cellular image segmentation for RNAi genome-wide screening. In J. Duncan and G. Gering, editors, *Proc. 8th Internat. Conf. on Medical Image Computing and Computer-Assisted Intervention (MICCAI'2005)*, volume 3749 of *Lecture Notes in Computer Science*, pages 302–309, Palm Springs, CA, USA, Oct 2005. Springer-Verlag.

[202] C. Zimmer, E. Labruyere, V. Meas-Yedid, N. Guillen, and J.-C. Olivo-Marin. Segmentation and tracking of migrating cells in videomicroscopy with parametric active contours: a tool for cell-based drug testing. *IEEE Transactions on Medical Imaging*, 21(10):1212–1221, 2002.

[203] C. Zimmer and J.-C. Olivo-Marin. Coupled parametric active contours. *IEEE Transactions on Pattern Analysis and Machine Intelligence*, 27(11):1838–1842, 2005.